MW00563824

NOBEL PRIZES
and
Life Sciences

NOBEL PRIZES
and
Life Sciences

Erling Norrby

The Royal Swedish Academy of Sciences, Sweden

World Scientific

NEW JERSEY · LONDON · SINGAPORE · BEIJING · SHANGHAI · HONG KONG · TAIPEI · CHENNAI

Published by

World Scientific Publishing Co. Pte. Ltd.

5 Toh Tuck Link, Singapore 596224

USA office: 27 Warren Street, Suite 401-402, Hackensack, NJ 07601

UK office: 57 Shelton Street, Covent Garden, London WC2H 9HE

Library of Congress Cataloging-in-Publication Data
Norrby, Erling.
  Nobel prizes and life sciences / by Erling Norrby.
    p. cm.
  Includes bibliographical references and index.
  ISBN-13 978-981-4299-36-7 (hard cover : alk. paper)
  ISBN-13 978-981-4299-37-4 (pbk : alk. paper)
  1. Life sciences. 2. Nobel Prizes. I. Title.
  QH315.N667 2010
  570.79--dc22

                               2010031703

British Library Cataloguing-in-Publication Data
A catalogue record for this book is available from the British Library.

Image on page 39: *The Three Younger Sons of Shah Jahan*, Mughal painting by Balchand (India, early 17th Century). Courtesy of V&A Images/Victoria and Albert Museum, London.

Copyright © 2010 by Erling Norrby

All rights reserved.

Printed in Singapore by Mainland Press Pte Ltd.

Preface

This book project has evolved over a long time. During 1997–2003 when I was the Permanent Secretary of the Royal Swedish Academy of Sciences I frequently gave lectures around the world on "A century of Nobel Prizes" later with an added prefix "More than … ." This theme was used for my inaugural lecture at the American Philosophical Society and a short version of the text was published in the journal of the Society (Norrby, E., A century of Nobel Prizes. *Proc. Am. Philos. Soc.* 2002; 146:323–336). Chapter 1 is a markedly expanded version of this article. In particular I have researched Alfred Nobel's contacts with the Royal Swedish Academy of Sciences and with the Karolinska Institute. This was done in order to understand his motivation for the formulations used in his final will. The first chapter also reflects on the process of selection of Nobel Prize recipients and how this has evolved over time. These developments are further illustrated by the different examples of prizes given in later chapters.

Throughout the writing of this book I have used my own experiences of working with Nobel Prizes as a sounding board. These developed after I had become Professor and Chairman of Virology at the Karolinska Institute in 1972. For some 20 years, from 1973 onwards, I was involved in the committee for the prize in Physiology or Medicine in different functions mostly as an adjunct or ordinary member of the committee. The meetings with the committee were the best experiences during my 37 years at the Institute. Everyone on the committee was aware that we were carrying a unique responsibility. We were expected to try to understand all the advancing frontiers of the field of biomedical research and related fields of life sciences. Based on this insight we were to judge priorities, allocate these to the appropriate scientists and compare the relative magnitude of the discoveries proposed. These meetings fostered friendships transgressing the traditional if unfortunate territorial atmosphere of medical faculties. When I moved on to head the Royal Swedish Academy of Sciences in 1997, I brought with me valuable experiences of the Nobel work at the Karolinska Institute for my supervision of the work with the Nobel Prizes in Physics and Chemistry at the Academy, and in my function as a member of the Board of the Nobel Foundation.

In 2004 I was invited by one of the previous visiting Japanese scientists in my laboratory, Takehiro Togashi from Sapporo, Hokkaido, to give a lecture on "Serendipity and Nobel Prizes." This was an interesting theme to review and it gave me insight into the life of a word from its first introduction. Serendipity has become a popular word in science. It is striking how incidental events may cause unexpected reorientations of the course of science. In fact the experience is that many of the discoveries that have had a major impact on the course of societal developments were not intended, at the time they were made, to provide the opportunities they later were found to offer. The material collected for this lecture was used in the writing of Chapter 2. In some parts of this chapter I have relied extensively on documentation in the book *The Travels and Adventures of Serendipity* by Robert Merton and Elinor Barber. Harriet Zuckerman gave valuable advice for this chapter and also provided important interview material for Chapters 4 and 5. She also made available the picture of Merton, her late husband.

In 2005 I had a meeting with Stanley Prusiner (presented in Chapter 8 on which he also gave good advice), a friend for many years and Nobel Laureate in Physiology or Medicine in 1997. We discovered that we were both interested in the events that led to the Nobel Prize in Physiology or Medicine to John F. Enders, Thomas H. Weller and Frederick C. Robbins for their discovery that polio virus could grow in cells of non-nervous origin. Because of their momentous discovery it became possible to develop effective vaccines and eliminate one of the major epidemic scourges of the first half of the 20th century in industrialized countries. The archives of relevance for this prize had just been released following the 50 years secrecy rule. We obtained permission from the committee for Physiology or Medicine, represented by its then secretary Hans Jörnvall, to review the relevant documents. It should be noted that there are strict rules controlling access to the Nobel archival material. They can be made available after a written application, but only for scholarly work. The documents were put at our disposal at the Nobel secretariat by the kind assistance of Ann-Margreth Jörnvall and Agneta Sjövall. They have continued to provide excellent assistance during my yearly January visits after 2005. In 2010, the new secretary of the committee, Göran Hansson, gave me permission to use the archives.

Since all the documents of the archives, except some of the nominations, were in Swedish I first had to translate them into English. Hereafter, Prusiner and I could sit down at the Nobel Forum to review them. We made exciting discoveries. I might add that the personalities we encountered in the documents were people that I had met as teachers at the Karolinska Institute or as colleagues during my early career as a virologist. One important person in the polio field was my predecessor as Professor of Virology, Sven Gard, who turned out to have had a major influence not only on the 1954 prize, but also on many other prizes during the 1950s and 1960s. It makes a major difference to write about people, whom one has met and sometimes even learnt to know quite well.

As we were leaving the Nobel Forum after one of our sessions we happened to meet a virology colleague Lennart Philipson. We told him about our project and he then mentioned that in 1954 there was an exceptional leak about discussions concerning the prize in Physiology or Medicine to *The New York Times*, information we could follow up. We eventually summarized our findings in a manuscript which was published in a regular scientific journal (Norrby, E. and Prusiner, S. B., Polio and Nobel Prizes: Looking back 50 years. *Ann. Neurol.* 2007; 61:385–395). This publication provided the core part of Chapter 5, which includes some supplementary material and adjustments to harmonize with other chapters. Many individuals were helpful in the development of the article and hence also this chapter. Not only did we have permission to use the archives at the Karolinska Institute, but also at the Royal Swedish Academy of Sciences. Columbia University Oral History Research Office gave us access to Robbins's reminiscences and our colleagues Margareta Böttiger, Hilary Koprowski, Erik Lycke, Michael Katz, Samuel Katz, Neal Nathanson, Weller himself (still alive at the time) and Julius Youngner gave valuable assistance.

The interesting experience of reviewing the archives of the 1954 prize inspired me to review the 1951 prize to Max Theiler. Again the archives revealed very interesting circumstances and eventually I wrote a comprehensive essay about the history of yellow fever and about the prize. Material for research was made available not only at the Karolinska Institute but also at the American Philosophical Society — where, however, Theiler was not a member — by kind assistance of Marty Levitt, and in particular at the Archives of the Rockefeller Foundation, Sleepy Hollow, New York, where I was well cared for by Lee Hilzik and Darwin Stapleton. Stapleton also generously edited the language of my essay and in addition provided valuable material for Chapter 6. Furthermore the Columbia University Oral History Office again provided help by making Theiler's reminiscences accessible. A number of colleagues read the text and gave valuable advice; Baruch Blumberg (presented in Chapter 2), Günter Blobel, Purnell Choppin — who himself had reviewed the history of yellow fever — and Rolf Luft. When the final version of the yellow fever essay was shared with Blobel at the Rockefeller University, he suggested that it might be published in *The Journal of Experimental Medicine*. Eventually an article about half the length of Chapter 4 was published after a careful editing by Jennifer Bell (Norrby, E., Yellow fever and Max Theiler: The only Nobel Prize for a virus vaccine. *J. Exp. Med.* 2007; 204: 2779–2784). This publication appeared 70 years after Theiler's original prize-awarded publications in the same journal.

My next project exploring the rich content of the Nobel archives was to review advances in my own field virology. I decided to look into ways in which the prizes might illustrate the major events in the maturation of the virus concept. Archives both at the Karolinska Institute and not least at the Center for the History of Sciences at the Royal Swedish Academy of Sciences, where I have my office, were extensively used. My manuscript was reviewed by Marc H. V. van Regenmortel, Marian Horzinek and Frederick A. Murphy, who gave valuable advice and eventually it was published in a regular scientific journal (Norrby, E., Nobel Prizes and the emerging virus concept, *Arch. Virol.* 2008; 153: 1109–1123). This article with some modifications came to be Chapter 3.

There were peculiarities with both the 1951 and 1954 prizes in Physiology or Medicine. In 1951 there was no external nomination of Theiler, who was therefore proposed by the chairman of the committee, Hilding Bergstrand, and in 1954 Weller and Robbins were nominated for the first time in 1954, the same year in which they were awarded a prize. It seemed that in both these cases the committees were eager to award the prizes and I became curious to know how often similar circumstances had prevailed previously. At this phase of the work it was possible to review the archives of the first 50 out of the total of 100 Nobel Prizes in Physiology or Medicine that have been awarded by 2009. I gave the essay the rubric "Unusual Nobel Prizes," but it turned out that they were not that unusual. There were a total of nine prizes, including 12 prize recipients, who became awarded the same year that they were first nominated — including 1901, when by definition this had to be the case — and there were five prizes, involving six prize recipients, who received a prize without any external nomination. This led to a long essay covering many different fields of biomedicine and it became Chapter 6. I am grateful to a large number of colleagues for advice on different parts of this comprehensive essay; Michael Bliss, Derek Denton, Georg Klein — an invaluable source of information also for Chapters 7 and 8 —, Ulf Ragnarsson and Bengt Saltin.

When this essay was finished the idea emerged that a book containing the different materials for the lectures and for the essays might be developed. I put this idea to Professor Kok Khoo Phua in Singapore in February 2009. His immediate response was that he would be happy to publish the material through his company World Scientific Publishing, which had already produced a large number of books related to the Nobel Prizes.

In order to form a bridge to present day science, which had already been attempted to varying degrees in Chapters 3–6, I decided to add two chapters.

It had become clear already from my work on Chapter 3 that one of the most dramatic revolutions in scientific comprehension was the eventual insight that DNA represented the genetic material. This started to emerge already in the mid 1940s thanks to the epoch-making work of Oswald T. Avery at the Rockefeller Institute. Eventually my research into the area and into relevant Nobel archives, including the background documents relating to prizes in Chemistry to Alexander R. Todd in 1957, in Physiology or Medicine in 1958 to George W. Beadle, Edward L. Tatum and Joshua Lederberg, and in 1959 to Severo Ochoa and Arthur Kornberg, allowed me to write Chapter 7. However, this chapter does not end in 1959; it also discusses later developments in the rapidly growing field of molecular biology. It reflects in particular on how the emphasis on nucleic acids and proteins as the hegemonic molecules of life has shifted over time. Hence it was given the subtitle *A drama in five acts*. Few fields have been so well reviewed as our fascinating growing insight into the structure and function of the genetic material and my main emphasis therefore is on the shifts in opinions and interpretations of the field. For some years I have been the Vice Chairman of the Board of Trustees of the J. Craig Venter Institute, a highly dynamic private research enterprise located in Rockville, Maryland and La Jolla, California. My contacts with scientists at this Institute and other members of the Board have provided important insights into the rapidly moving field of molecular genetics.

The last chapter of the book is different. No Nobel archival material was used in its preparation. It discusses a field of research with which I am personally well acquainted and where I have befriended the two main actors. It was particularly delicate to write since I also wanted to reflect on personality traits of successful scientists. The field includes some of the most unexpected turns in our comprehension of knowledge, in this case the identification of a remarkable new form of infectious agents. Insights into their mechanisms of causing tissue damage have led to the identification of a previously unrecognized category of protein aggregation diseases of great relevance to human medicine. It has also brought to light a new field of previously unknown information exchanges between protein molecules. Many people have provided material and given advice of importance for the structure of Chapter 8; Annica Arnberg and collaborators, Paul Brown, Bruce Chesebro, David Eisenberg, Susan Lindquist, Dorrie Runman — Carleton Gajdusek's daughter-in-law — and Rolf Seljelid. Runman gave particularly valuable information about the Gajdusek household and provided useful pictures.

The pictures used in the book are drawn from many sources, but I would like to particularly thank Jonna Persson at the Nobel Foundation and, not mentioned above, Richard Krause, Ulf Lagerkvist, Jan Lindsten, Nils Ringertz, Lennart Stjärne and Hamilton Smith. Since English is not my native tongue I have had all the chapters in their final form read by Harry D. Watson to secure the use of idiomatic language.

Colleagues at the Center for the History of Science have contributed to this book in many ways. Its Director Karl Grandin gave me permission to examine the Nobel archives of the Academy and spent many hours with me working on illustrations. Maria Asp, Anne Miche de Malleray, Jonas Häggblom and Bengt Jangfeldt — a richly endowed humanist and successful author — have provided joy, inspiration and invaluable intellectual as well as practical support.

This book is dedicated to Margareta, my wife — and love — for more than 50 years. Without her unfailing support it is questionable if it would ever have been finished.

Contents

Chapter 3

Nobel Prizes and the Emerging Virus Concept 66

Chapter 4

The Only Nobel Prize for a Virus Vaccine:
Yellow Fever and Max Theiler 97

Chapter 7

Nobel Prizes and Nucleic Acids: A Drama in Five Acts 195

Chapter 8

Nobel Prizes, Prions and Personalities 245

Chapter 1

More than a Century of Nobel Prizes

The five Nobel Prizes have evolved to become unique institutions. At the time when they were conceived the awards were the largest ever and they were international. Over the years these prizes have acquired the status of being an exceptional measure of achievements in human endeavors, not the least the remarkable discoveries in the fields of natural sciences.

The first prizes were awarded in 1901 and, except for some years during the first and the second world wars and a few, variable number of, additional years, they have been distributed annually. In 2009 the total number of prizes that have been awarded in Physics, Chemistry, Physiology or Medicine, Literature and Peace are 103, 101, 100, 102 and 90, respectively. With this long-term perspective it is attractive to reflect on what it is that confers on them their unique prestige and what can be learnt from reviewing the selection process, the discoveries that have been awarded and the anointed scientists, authors and world citizens. Can we possibly get some insight into the remarkable creative processes that foster the progress of knowledge and further cultural developments that lead to maturation in the realms of esthetics and human interactions? Have the prizes fulfilled the laudable intention of recognizing advances that are "to the benefit of mankind"?

This book discusses only the prizes in natural sciences reflecting the fields of which I have a wide personal experience. Information about the prizes in Literature and in Peace can be found in separate books[1-3].

Alfred Nobel and his life

There is a considerable literature on Alfred Nobel and his prizes[4-6] and about the Nobel family[7]. The following is a telescoped presentation of the eventful, mostly vagabond, international life of the shy and very private person who was Alfred Nobel. He was born in 1833 in Stockholm and was the third of four surviving brothers out of originally six children. Two centuries back his ancestors were farmers in the southern part of Sweden. Since they resided in the parish of Nöbbelöv they took the family name Nobelius, the Latinized derivation of the name of this village. In genealogical presentations it is often pointed out that Alfred's grandfather, four generations back, Petrus Olai Nobelius married a daughter of a very renowned Uppsala polymath, Professor Olof Rudbeck, a central figure at the university of that city.

Alfred's dynamic father Immanuel was a self-taught engineer and businessman with highly variable successes in his ventures. At the year of Alfred's birth, the father was forced into bankruptcy of his architect and building activities in Stockholm. The family had to manage on relatively limited resources and through these challenging times the initiatives taken by his wife Andrietta were invaluable. She is described as a richly endowed woman showing intelligence, humor and a warm heart. When times were difficult she always managed to find ways forward for her family. All the boys, not the least Alfred, came to respect and admire her. Her influence on their future courses in life should not be underestimated.

After some lean years in Stockholm, Immanuel in 1838 moved to St. Petersburg in Russia to try his luck there. With time his business began to flourish, partly because he was successful in establishing contacts with the Russian government. These contacts became particularly valuable at a later time when Russia in 1853–1856 became involved in the Crimean war. Four years after his arrival in St. Petersburg he could eventually move his whole family there. Until then Andrietta had supported her family by running a little store with dairy products and some greens. At the time of movement Alfred was nine years old and had finished the single year of formal schooling that he came to get in his life. He received his further education by private tutors. The older brothers were trained as engineers while Alfred got a training in chemistry. He was quick in learning and managed five languages fluently when he was 17 years old. His personality has been described as giving "the impression of a prematurely developed, unusually intelligent, but sickly, dreamy and introspective youth, who preferred to be alone"[5]. In

Alfred Nobel at the age of about 30 years. [Courtesy of the Nobel Foundation.]

his late teenage years, in 1850, he was sent abroad for two years to Germany, France, Italy and North America. For a while he was the assistant of the famous Swedish inventor John Ericsson in New York. In Paris he worked in the renowned laboratory of professor Théophile-Jules Pelouze, who had a wide international network and had been in correspondence with the highly respected Swedish biochemist Jöns Jacob Berzelius. In Pelouze's laboratory Nobel met a young Italian scientist Arcanio Sobrero, who in 1847 had discovered an explosive oil that he named nitroglycerin. Alfred saw the potential of this explosive and developed a design for its practical use for which he got a patent in 1863. The product was later named dynamite. At his death 33 years later there were 355 patents registered in his name.

In 1863 the family, except for the two older brothers Robert and Ludvig, were back in Stockholm after the father had been forced into another bankruptcy of his for a while very successful business in St. Petersburg. When the Crimean war over his favorable dealings with the Russian Government had started to peter out, Alfred then took the initiative, together with his father, to start a factory at Heleneborg in Vinterviken at the outskirts of Stockholm for the production of explosives. Some time after this the family was hit by another tragedy. A major explosion occurred in a storehouse killing the youngest Nobel brother, Emil, recently registered as a student at Uppsala University, and four other people. This was a severe blow to the dynamic father of the family and certainly contributed to his progressive decline and death in 1872.

Alfred, however, advanced towards an impressive career as an international industrialist and inventor. His inventions facilitated the development of the new world of communications. Roads, tunnels and channels were built by use of his explosives. He established factories in many countries, but preferred to live in France from the age of 40. In 1873 he bought a house in Paris at 59 Avenue Malakoff, where he had his laboratory and office as well as his home (page 40). The laboratory activities were moved to a property in Sevan-Livry outside Paris in 1881. In his later years he also bought two other homes, a beautiful villa in San Remo in Italy, to which he in 1890 moved his experimental activities from France, and a mansion at Björkborn in Sweden, but he never truly settled and his whole life continued to be one of restless activities. There were periods when he was highly successful in his businesses, but there were other times when he failed and even came close to bankruptcy. As previously mentioned, he was a shy and reclusive person and he never married. Judging from his writings and his letters his views on life were generally very dark and pessimistic.

Academies and prizes

To bestow honors is a tradition in academies and learned societies[8]. The rewards can be of different kinds, medals to recognize important advances and prizes and grants to provide, in addition, financial resources for the pursuit of science. A clear distinction between the latter two forms of rewards did not develop until during the twentieth century. Since its beginning in 1739, the Royal Swedish Academy of Sciences has been involved in instituting prizes and distributing medals. In February 1868 Clemens Ullgren, a chemist and member of the Academy, proposed to the committee of the Letterstedt's prize at the Academy that it should recognize the father and son, Immanuel and Alfred Nobel, for their discovery of the new use of nitroglycerin, in the form of dynamite, for blasting. This also became the decision of the Academy soon thereafter. The prize recipients could choose between a gold medal and a cash award of 996 Swedish crowns (Riksdaler). They accepted the medal and the father took care of it. After the death of his mother, the gold medal became Alfred's property. The prize certainly served as an official recognition and encouragement, not the least to the then severely ailing father. To Alfred Nobel it meant a first contact with the Academy. Over the years he came to interact with some members of this institution, in particular the explorer Adolf Nordenskiöld. They were in correspondence from 1874 until Alfred Nobel's death.

The Royal Swedish Academy of Sciences operates like Plato's classical Academy (the name derives from the olive groove, *hekademia*, where it resided) in old Greece. It has a fixed number of national and international members and a new member is elected when an empty slot becomes available. Originally this happened when a member died, but since some fifty years a member who becomes 65 years old is no longer included in the fixed number, although he retains all the rights as a member. The members of the Academy are divided into classes for mathematics, physics etc. with a fixed number of members in each class. The system for division of members into different classes has changed with time. Proposals for a new member can only be made by members of the Academy.

In 1884 the Academy had a possibility to elect a new foreign member to its class of economical sciences. The mining engineer Anton Sjögren proposed Alfred Nobel for a membership. He referred to the fact that Nobel "… had introduced the technical professions to the use of nitroglycerin in the manageable forms of dynamite and derived preparations …" and that this "… has contributed to the progress of blasting of rocks more than any other discovery made during this century … ." The nomination concluded in the spirit of the time by stating that Alfred Nobel had "… lent luster to not only his ancestral home but also his native land." At its *in pleno* meeting on March 12 the Academy elected Alfred Nobel as a member. Although he was chosen as a member of the class of economic sciences, it was his contributions as an engineer that was highlighted, not his success as a businessman. It should be noted that at the time the term economic sciences had a much broader meaning than it has today. Referring to the original etymological derivation of the word economy, from Greek *oikos* and — *nómos* meaning home and the one who manages it, respectively, it referred to societal housekeeping at large. On the occasion of the revision of the class system of the Academy in 1842 the opinion had been expressed that its class for economic sciences should include also technical and statistical sciences. In a later revision in 1904, this original class was split into two classes, one for technical sciences and another for economic, statistical and social sciences.

Nobel was informed by his Academy contact, Nordenskiöld, who in fact was not present at the meeting when the election had occurred, by telegram about his new distinction. Nobel promptly responded by a letter:

> Brother! Having just returned from a short business trip I was surprised by an equally unexpected and pleasant telegram from you. The high distinction, for which I solely have to be grateful to your influential kindness, makes me blush, when recognizing how undeserving I am. I view the recommendation that you and your colleagues have given, not as a reward for the limited contributions I may have made, but as an encouragement for forthcoming activities. If I, spurred by such an encouragement, do not succeed in being of some value in the realms of progress I shall bury my wretched soul alive in some isolated, unrecognized corner of the world. But my deep gratitude to my great countryman's generosity and kindness will follow me all the way to that place. With sincere devotion, A. Nobel.

The attention given by the Academy apparently moved and inspired Nobel and it is not unlikely that it influenced his eventual decision to bestow on it a major responsibility for his prizes to be. However, as far as is known, Nobel never visited the Academy and he never developed any technical or scientific collaboration with any of the Academy members. He seems to have had, in particular in the 1890s, closer contacts with representatives for another of the forthcoming prize-awarding institutions, the Karolinska Institute.

Alfred Nobel's mother, Andrietta. [Oil painting by Anders Zorn. Courtesy of the Nobel Foundation.]

Over the years Alfred kept in close contact with his beloved mother. He wrote letters to her regularly and always remembered her on her birthday. If possible he tried to visit her in Sweden on these occasions. In 1889 Andrietta eventually died, aged 84 years old. Her economy had gradually stabilized, but she had continued to live under relatively unpretentious conditions. Alfred Nobel inherited one-third of the relatively large sum of money she left behind. He used a minor fraction for the decoration of her tombstone, but the major part of the money, 50,000 Swedish crowns, was donated to the Karolinska Institute for the establishment of Caroline Andrietta Nobel's fund for the support of medical research "... within all branches of medical sciences and the use of the research achievements for education and for literature".

In the discussions about the donation in his mother's name Nobel learnt to know Sven von Hofsten, a pediatrician and associate professor at the Institute[9]. He asked von Hofsten for advice regarding the possible availability of some young promising physiologist that could assist him in experiments he wanted to initiate in Paris. von Hofsten recommended a younger colleague, Johan (Jöns) Erik Johansson (page 145), who at that time, in 1890, was visiting Leipzig. He had just defended his academic thesis in physiology at Uppsala University and had had his teaching competence (docentship) acknowledged at the Institute in Stockholm. After letter contacts with Nobel, Johansson traveled to Paris in October 1890 and stayed for five months. During this time he performed some studies concerning blood transfusions. Nobel was so inspired by the contacts that he tried to engage Johansson in the establishment of a new institute in Paris to perform different new kinds of experimental medical research.

It should be emphasized that Nobel had his own, rather critical view of the medical profession. He referred to physicians as "faculty donkeys." By some irony of fate Nobel was recommended his own "blasting oil" nitroglycerin to alleviate his circulatory heart problems that started to develop in 1890. However, he did not take this remedy and wrote to his brother Robert, "They call it trinitrin in order not to scare pharmacists and the public."

Nobel had new ideas about compounds to be used for destruction of bacteria and speculated about new ways of administering anesthetics. Furthermore he was particularly interested in the analysis of the chemistry of blood as a means to make diagnoses of different diseases and in techniques for blood transfusions. Previously blood transfusions had been tried as an indirect method, but Nobel's idea was to directly transfuse blood from one individual to another. This eventually turned out to be more than a technical problem as Karl Landsteiner from Austria later demonstrated. In 1930 he received a Nobel Prize in Physiology or Medicine for his demonstration of the different blood groups in humans. In spite of the tempting offer Johansson eventually decided not to stay in Paris. He returned to Stockholm to pursue his career at the Karolinska Institute. Still he retained a good contact with Nobel and came to play a major role in the future developments concerning the prize in Physiology or Medicine as described in Chapter 6.

During an academic peregrination, Axel Key, the Vice-Chancellor (Rector) of the Karolinska Institute, in 1893 visited Alfred Nobel in San Remo. Key

was treated to a splendid dinner at the magnificent house. During the meal he took the opportunity to praise the donation that Nobel previously had made in the name of his mother. Nobel's response, according to a letter Key sent to his wife, was that "this fund is not the last grant with which Nobel intended to endow the Institute. He had made further provisions in his will." It seems that during the 1990s Nobel's interest in experimental medicine grew. He donated a considerable sum to support the research carried out by Ivan Pavlov, the future Nobel Laureate in Physiology or Medicine, and his colleague Marceli Nencki in St. Petersburg. In 1895 Nobel sent his young private assistant, the young engineer Ragnar Sohlman to Stockholm to supervise chemical analyses of urine samples from patients with various diseases collected at St. Göran's Hospital in the city. It was Key who had arranged for the analyses of these samples at the Karolinska Institute.

Axel Key (1833–1901), Vice-Chancellor of the Karolinska Institute (1886–1897). [Courtesy of the Karolinska Institute.]

In 1893 Nobel was honored with another recognition by the academic establishment in Sweden. He was selected to become an honorary doctor of philosophy at Uppsala University in connection with its 300 years jubilee promotion celebrations. This university is the oldest in the Nordic countries, having been founded in 1477. However, during the major part of the

16th century, it was dormant to be reopened in 1593, hence the jubilee. It was requested that the honorary doctors provide a brief autobiography. Nobel wrote:

> The undersigned is born on October 21, 1833, has acquired his knowledge by private education without attending an institution for higher learning (*högre läroverk*); has been active within the field of applied chemistry and developed explosives, recognized by the names dynamite and blasting dough and smokeless gunpowder, also named ballistit and C.89. (I) am a member of the Royal Swedish Academy of Sciences, of the Royal Institution (Society?) in London and the Société des Ingenieurs Civils in Paris. (I) am since 1880 a Knight of the Order of the Northern Star. (I) have the degree of Officer of the legion of Honor. Printed documents: only a lecture in English, which was awarded a silver medal.

Although Nobel never showed any esteem for formal honors it is clear from his personal correspondence that he valued the attention he was given by the Swedish academic establishment and the Government. There is no archival material documenting the reasons for giving Nobel the Order of the Northern Star. The only remaining documentation is a receipt showing the Nobel had made the requested payment in Paris for the Order, 64 Swedish Crowns.

In 1893 Nobel was approaching the end of his life and he naturally reflected on what would happen to his wealth after his death. Because of the lonely life he had chosen he did not have any immediate heirs. There were several steps before he formulated his famous final will in November 1895.

The will

Alfred Nobel's last will was written in Swedish and deposited in a Swedish bank. He wrote it without any legal assistance. In fact due to some disappointing experiences he had had in defending his patent rights he referred to lawyers as "parasites exploiting legal formalities." Nobel wrote his first will in the early 1890s, but this will has not been found. It was probably destroyed. In the next version of his will, from 1893, he did not specify exact amounts, but made the provision that 20% of his capital should be distributed among 22 named relatives, collaborators and friends. A further 16% should be given to various institutions. Among these was "*Oesterreichische Gesellschaft der Friedensfreunde*" in Vienna, a society for peace founded by Bertha von Suttner, the single friend who inspired him the most to develop pacifistic engagements. Other institutions mentioned were the Swedish Club in Paris, Stockholm's *Högskola* — at the time an embryonic private institution of learning, which later became a Governmental university — , a private hospital in Stockholm and the Karolinska Institute. The remaining 64% of the capital should be used to establish a fund at the Royal Swedish Academy of Sciences for the awarding of yearly prizes for the most pioneering discoveries and theoretical works within the field of science exempting the domains of physiology or medicine. In a somewhat puzzling way Nobel specifies in association with this bequest that "Without making this an absolute condition it is my wish that those who by their writings or actions have been most successful in fighting prejudices, which are still held by nations and governments against the establishment of a European peace tribunal, should be considered. It is my distinct wish that all in this will defined prize awards should be given without any consideration to whether the recipient is Swedish or foreigner, man or woman." The advantage of this will over the final one was that it clearly specified the recipients of the money. Still it would have been very difficult for the Academy to define the extensions of its missions and fulfill them. How would it engage in awarding good literature and in furthering peace efforts?

The portal paragraph in the final will is that prizes should be awarded in five areas "to those who during the preceding year have done the greatest benefit to mankind." These are the words of an idealist and perhaps a dreamer. It is tempting to deduce that Nobel had acquired this altruistic philanthropic attitude from his mother. It seems that she had been successful in instilling in all her three surviving sons a respect and responsibility for fellow human beings. All of them became successful businessmen. The oldest, Robert, became a leading developer of the petroleum industry in Baku, and the second one, Ludvig, became the founder of a very prosperous arms factory in St. Petersburg and was also involved financially in the Baku projects[10]. As was apparent when the old archives became available after the fall of the Soviet Union in 1991, the two Nobel brothers were pioneers in the way they cared for their employees and attempted to improve their conditions of work. The regulated working hours were reduced and schools, hospitals, libraries and living quarters were built. Ludvig achieved an outstanding position in the Russian industry and when he died in 1888 this was noted widely. A number of scholarship funds were established and a science prize was established by Nafta Company Brothers Nobel. A qualified jury selected the prize recipients and it was awarded three years 1896–98. According to Bengt Jangfeldt[10] it can even be seen as the first "Nobel

Prize." Alfred Nobel only visited St. Petersburg once during his adult life, in March 1883. However, he had major financial interests in the Baku petroleum operations. He was one of the founding brothers when it started in 1878. Stocks in this company represented the largest single asset in the fortune that he left behind, 7.5 out of the total of 33 million Swedish crowns.

Alfred Nobel had his final will witnessed at the Swedish Club in Paris before it was deposited in the Swedish bank. Its content was kept highly secret. The handwritten document prepared by Nobel himself had a number of formal defects, which led to a series of complications before the will could eventually be implemented. When the will was opened five days after Nobel's death at the age of 63 on December 10, 1896, the relatives learnt to their dismay that only a limited portion of the estate was bequeathed to them. Only about one million Swedish crowns, as compared to more than three million Swedish crowns in the previous, canceled will, was given to them, primarily to his three nephews. The rest of the fortune should be used for five prizes. Why then was it Nobel's wish that his estate should be used for prizes? His idealistic and altruistic attitudes to life has already been mentioned, but additional factors could be that his political views had a socialistic color and that he did not sympathize with transfer of wealth between generations. Since he himself was a true inventor he could appreciate the importance of providing creative conditions for young talented inventors. His concept was simple. The prizes to be given should allow the awardees to concentrate on their work without any need for income for some 20 years.

The will specifies five different prizes. The first three are in natural sciences and are given to the one (a) who in the field of physics has made the most important *discovery* or *invention*, (b) who in the field of chemistry has made the most important *discovery* or *improvement,* and (c) who has made the most important *discovery* in the domain of physiology or medicine. The common denominator for these three prizes is *discovery*. Thus, prizes are not given for life contributions to science but, in most cases, for the making of a single discovery with a huge impact. Such discoveries mostly emerge from experimental studies, but also major theoretical contributions have been awarded. In physics it took until 1922 before such theoretical discoveries were awarded. In this year Albert Einstein got the prize for 1921 "for his services to Theoretical Physics, and especially for his discovery of the law of the photoelectric effect (not his discovery of the theory of relativity; my remark)" and Niels Bohr the prize for 1922 "for his services in the investigation of the structure of atoms and of the radiation emanating from them." Only in some few cases is it possible to identify prizes in which a particular reference to the word "invention" and in particular "improvement" has been used. Still, in the early years, prizes in Physics were given for inventions as in the cases of Lippmann in 1908 for the color photography technique, of Marconi and Braun in 1909 for radio transmission and of Dalén in 1912 for automatic regulators in lighthouses. As late as 2009 half a prize was given to Charles Kao, "for groundbreaking achievements concerning the transmission of light in fibers for optical communication," and to Willard Boyle and George Smith, "for the invention of an imaging semiconductor circuit — the CCD sensor."

The two additional prizes, for Literature and for Peace, are given to the person who in literature has produced the most outstanding contribution *with an idealistic orientation,* and to the person who has worked the most or the best for fraternization between peoples and elimination, or reduction of standing armies and formation and dissemination of peace congresses.

The prize-awarding institutions are, for physics and chemistry, the Royal Swedish Academy of Sciences, for physiology or medicine, the Karolinska Institute, and for literature, the Swedish Academy. All these institutions have a long history. The Royal Swedish Academy of Sciences was established in 1739, originally to further the development of useful knowledge and more recently "to further the advance of science, in particular natural sciences and mathematics." The Swedish Academy was founded in 1786 "to advance the Swedish language and Swedish literature." The youngest of these three institutions, the Karolinska Institute, was founded in 1810 and is today a medical university and a dominating center of learning in medicine and life sciences in Sweden. The Peace prize is a responsibility of the Norwegian House of Parliament (*Stortinget*), which selects a five-member committee carrying the full responsibility for selecting prize recipients. In this context it should be mentioned that Sweden and Norway formed a union until 1905, when it was peacefully resolved. However, engagements in Nobel Prizes have remained a continued shared responsibility of the two countries. Still it took until the 1970s before the Norwegian Peace prize organization became represented in the board of the Nobel Foundation.

Nobel's choice of the five fields has been a matter of many discussions. As can be seen from his second, canceled will he originally considered to bequest his estate, in addition to money designated to relatives and other heirs, to a number of specified institutions. The newly established private center of higher learning, Stockholm's *Högskola,* was among the latter. Certain frictions with the colorful Vice-Chancellor of this institution in 1891–1892, the mathematician Gösta Mittag-Leffler, may have made him change his mind and perhaps they also made him exclude mathematics, the "mother of all sciences." However Mittag-Leffler himself bequeathed his estate to the Royal Swedish Academy of Sciences for the establishment of a separate institute, which today represents a world famous institution for mathematics under its aegis.

The inclusion of literature may reflect Nobel's own engagement in this field. He made some unsuccessful attempts to write poems and fiction. Towards the end of his life he also wrote a drama entitled "Nemesis," which he had printed at his own cost. After his death the relatives made sure that all copies of the book except one were destroyed. A dramatized form of his play was recently shown in Stockholm. Perhaps the suffix "with idealistic orientation" — the meaning of which has been very difficult to interpret[1] — reflects his belief that literature is a means for changing our world. The idealistic focus is apparent already in the beginning of the will in which it says that awardees should "have made the greatest contribution to mankind," as already mentioned. His idealism, furthermore, is obvious also from his inclusion of a prize for peace efforts. In fact it is said that he believed that his development of the most effective explosives of his time would discourage violent interactions between groups of peoples, a somewhat naïve idea, as proven by the continued historical developments. In his deliberations about a possible Peace prize it is likely that he was inspired by his acquaintance with the Austrian pacifist Bertha von Suttner, who for a short while was employed as his private secretary. In 1905 she herself received the Peace prize.

Implementation of the will

There were many roadblocks to be removed before the final will of Alfred Nobel could be fully implemented. There were legal formalities such as the jurisdiction over the will. Furthermore there existed no legal and organizational structure to take responsibility for the fund. Nobel's relatives living in Sweden wanted to take advantage of the situation of uncertainty and contested the will in 1898. However Nobel's nephew Emanuel, representing the Russian branch of the family, supported a resolution in accordance with his uncle's will. In fact he even had an argument with the Swedish King Oscar II about the interpretation of it. The King, in keeping with the prevailing mood of the time of national chauvinism, did not like the formulation that "no consideration whatever shall be given to the nationality of the candidates, but that the most worthy shall receive the prize, whether he is Scandinavian or not." He did not join in the first prize-awarding ceremony in Stockholm. However, at all later events the prizes have been handed over by the Swedish King or occasionally by the Crown Prince. At the prize ceremony in Oslo, the Norwegian King honors the ceremony by his presence, but the Peace prize is handed over by the chairman of the committee.

Eventually a settlement was made with the relatives, who received a minor compensation. Emanuel, who was a highly successful businessman but not an imaginative inventor like some of his ancestors, was recognized on two occasions for his critical contributions to the establishment of the five prizes his uncle had specified in his will.

Emanuel Nobel (1859–1932), Alfred Nobel's nephew. [Courtesy of the Nobel Foundation.]

In 1910 he became an honorary doctor at the Karolinska Institute and a year later he was elected a foreign member of the Royal Swedish Academy of Sciences. The ceremony at which Emanuel received his honorary doctor's degree, his *doctor honoris causa*, at the Karolinska Institute was a special event. In 1910 the Institute celebrated its centenary. It was established after the last war that Sweden fought, the war with Russia when Sweden lost Finland. Even though the Institute had existed for one hundred years it had had a relatively slow development. There had been extensive and intense fights with the much older Uppsala University about who would have the right to give degrees. In fact it was not until 1910 that the Institute for the first time had the right to confer honorary doctor's degrees. The celebration was a splendid event in the great hall of the House of Nobilities on December 13 in the presence of the Royal Family and other exquisite guests. Among these was the 1910 Nobel Laureate in Physiology or Medicine, Albrecht Kossel from Germany (Chapter 7), who had received his prize three days earlier. Besides Emanuel there was one more foreign honorary doctor. This was the 1902 Nobel Laureate in Physiology or Medicine, Ronald Ross, the pioneering malaria researcher. In addition to the two foreign honorary doctors, the same degree was bestowed on eight Swedish physicians, and ten more physicians received their doctor's degree after having officially defended their thesis. Among the latter were several personalities which will appear in later chapters, like Erik Ahlström, Folke Henschen, and Carl Kling.

Emanuel Nobel's election to the Royal Swedish Academy of Sciences was based on a proposal by Gustaf Tamm. In a letter to the class for economics, statistics and social sciences he gives four reasons why Emanuel should become a member of the Academy. First and foremost he refers to the fact that Nobel has made outstanding work for the practical implementation of economic sciences and in particular for influencing the disposition of the assets defined in Alfred Nobel's will. This contribution had come to be of great value for the support of sciences by the Academy. In addition he refers to

Ragnar Sohlman (1870–1948), Alfred Nobel's close collaborator and executor of his will. Executive Director of the Nobel Foundation (1929–1946). [Courtesy of the Nobel Foundation.]

the documented success of the Nobel's nafta production in Russia, to the fact that Emanuel has become honorary doctor at the Karolinska Institute and that he had given 15,000 Swedish crowns to a research expedition. The proposal was accepted by the class and was sent to the Academy. It was then reviewed at two general meetings of this institution, and on November 11, 1911 Emanuel Nobel was elected a foreign member.

In spite of the help the executors of the will received from Emanuel Nobel it took some time before the proposed prize-awarding institutions, after their initial hesitance — in particular by the Royal Swedish Academy of Sciences — could be convinced to shoulder their tasks. The critical resolution of matters was the establishment of the Nobel

Foundation, an idea conceived by Ragnar Sohlman, one of the two executors of the will. This Foundation, which is not specified in the will, was instituted in 1900. Sohlman continued to have different roles in the development of the Foundation until his death in 1948. In particular he was its Executive Director from 1929–1946. Since 1992 his grandson Michael Sohlman has carried this

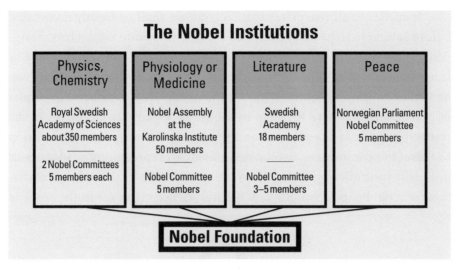

The Nobel Institutions

Physics, Chemistry	Physiology or Medicine	Literature	Peace
Royal Swedish Academy of Sciences about 350 members ——— 2 Nobel Committees 5 members each	Nobel Assembly at the Karolinska Institute 50 members ——— Nobel Committee 5 members	Swedish Academy 18 members ——— Nobel Committee 3–5 members	Norwegian Parliament Nobel Committee 5 members

Nobel Foundation

Figure 1. The prize-awarding institutions and the Nobel Foundation.

responsibility. The Nobel Foundation is an underlying, coordinating organization managing the fund, fulfilling legal functions and arranging the prize ceremony in Stockholm (Figure 1). However, it should be emphasized that it is the awarding institutions, not the Foundation, which carry the sole responsibility for selecting prize recipients and that give them their prizes. Sometimes the recipients are referred to as "winners," but the truth of the matter is that no one wins a Nobel Prize, he deserves it. The first Nobel Prizes could be awarded in 1901, as already mentioned.

The selection process

The organization of the prize work of the four prize-awarding institutions differs markedly. The prize work in Literature and Peace are in many regards unique and include certain particular inherent problems. In the Swedish Academy the five-member committee represents a sizable fraction of the total number of academicians. This number is 18, but in practice, because of aging and for other reasons, the number of available members is much lower. The Peace prize five-member committee only responds to itself, and therefore has a major exclusive mandate. Here I shall discuss the processes for selection of prize recipients in Physics, Chemistry and Physiology or Medicine, since these are the ones with which I have become personally acquainted.

It applies to all five prizes that a candidate shall be freshly nominated before January 31 to be eligible for a prize the same year. Only designated individuals have the right to nominate and the nomination shall be submitted in writing. Institutions cannot make nominations, but they can be invited to initiate nominations via their individual members. Examples of individuals who have the right to submit proposals are professors in the respective fields in Scandinavia and previous prize recipients. In addition to these, the committees also invite individuals representing academies or university institutions globally on a rotating basis. The total number of invitations sent out by the committees of natural science prizes are in the range of

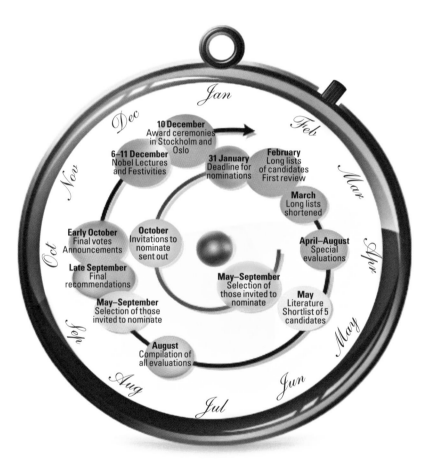

Figure 2. The Nobel year. The process of selection of Nobel Prize recipients is initiated as early as May–September the preceding year by listing the persons to be invited to nominate candidates. During the year when the prizes are awarded there are three particularly critical dates (weeks); January 31 — the last day of nomination; the first full week in October — announcement of the prize recipients in different fields on different days; and December 10 (Alfred Nobel's death day) — prize award ceremonies in Stockholm and Oslo.

2000–3000 and the number of nominations received have been between 200 and 500 during later decades. Roughly 10–20 percent of the nominations are new to the committee. The invitations are sent during the fall of the preceding year, starting the yearly cycle of events (Figure 2).

The present rules specify that there should be a committee composed of five members elected for a time period of three years. At the Royal Swedish Academy of Sciences the members may be re-elected twice but at the Karolinska Institute only once. In rare cases members may serve for an additional period after they have been out of the committee for a while. The working committee can be enlarged by electing adjunct members on a one-year basis. At the Academy there are only a small number of such members. The reason is that the committees of this institution interact with the class of Swedish members representing the field, which has about 35–40 members. Thus the committees do all the preparatory work leading to a specific proposal for prize recipient(s) in a field in a certain year. This proposal is submitted to the class which reviews it before it is eventually submitted to the meeting to which all members of the Academy, currently about 350 people, are invited. At these meetings, which generally gather more than 100 members, the final decisions are taken. The meeting starts with a general comprehensive review by the chairman of the committee of the moving frontiers and prize-worthy candidates in the whole field of physics or chemistry. Hereafter an in-depth presentation of the proposed prize recipients and their contributions is given, generally by another member of the committee. The proposal is then discussed and finally voted on. Academy members participating in these meetings enjoy the privilege of a unique insight into advancing sciences. There is usually an interval of about an hour between the time when the decision has been taken and the time when the result is officially announced. During this hour the Permanent Secretary has the particular responsibility to try to contact the prize recipients by phone, which sometimes means a 3 am wake-up call to colleagues in California. The Physics and Chemistry prizes are currently announced on the Tuesday and Wednesday, respectively, of the first full week of October.

Whereas the Academy represents all scientists in Sweden, the Karolinska Institute includes one of the totally six faculties of medicine in Sweden. It is, however, responsible for about 50 percent of all biomedical research in the country. At the Institute the College of Professors with lifelong tenure, the faculty, originally took the decision about the prize. Because of changes in rules for obligatory disclosure of official documents of governmental institutions, which include the Karolinska Institute, a new legal entity, the Nobel Assembly, was established in the late 1970s. This Assembly has 50 members and continuously renews itself as members retire or leave the Institute. In contrast to the Academy, which works with a three tier system, the Institute has only two levels of decision-making, the committee and the Assembly. This is one reason why the Institute for many years has used a larger working committee, including ten adjunct members. Another reason is that there is a need for a broad representation to cover the wide field of biomedicine. The committee interacts with and reports to the Nobel Assembly, which takes the final decisions on the recipient of the prize in Physiology or Medicine, currently on the Monday of the same week when prizes in Physics and Chemistry are announced.

The committees' work to identify the strongest candidates and eventually selecting a single proposal includes many steps. All nominations are scrutinized by the committees but, as mentioned, about 80–90 percent of them have already been encountered in the work during preceding years. Those who are newly nominated may be selected in a limited number of cases for a separate review but generally a note to the protocol suffices at the first time of nomination. Reviews can vary in their comprehensiveness, at the Karolinska Institute referred to as preliminary or complete investigations. Strong candidates in a common field frequently are reviewed together. Members of the committee may be reviewers but a large number of outside reviewers of both national and (only during the latter 50 years) international origins are used. Towards the end of August all reviews are collected in a paper-bound form, which at the Karolinska Institute provides the background material for the final discussions by the Assembly of the proposal made by the committee. At the Academy the collection of reviews is supplemented with a written comparative evaluation by the committee of the strength of candidates in individual branches of the discipline and a distillation of the strongest candidates into a single proposal.

These yearly volumes represent a fantastic real time analysis of advances in science of potential historical relevance. The goal of the selection process for Nobel Prize recipients is to identify contributions representing milestones in the evolution of discoveries in science. The exceptional reputation of the prize is based on the fact that the selection of recipients during the preceding century has been, if not flawless, at least of a very high standard. It should be emphasized that the process is managed by individuals who all have their varying scope and depth of knowledge and in addition their idiosyncrasies concerning the impact of scientific advances. A number of examples of diverging opinions and the way committees have managed to resolve these will be given in the coming chapters. In spite of these obvious limitations in the management of the selection process an overwhelming majority of the selected awardees in natural sciences and their contributions have stood the test of time. It is this mere fact that gives the prize its extraordinary international prestige. The prize simply has come to reflect the history of modern science.

The division into the disciplines of physics, chemistry and physiology or medicine is not absolute. Many candidates are nominated for more than one prize, in particular both in Chemistry and Physiology or Medicine. At the present time a double nomination of the latter kind includes more than ten percent of the candidates. In order to manage candidates nominated in more than one discipline there are joint meetings between two committees. It is obvious that in some prize areas like molecular biology many early prizes were given in Physiology or Medicine, as for discoveries of the structure of DNA, the genetic code, reverse transcription and restriction enzymes, whereas some later discoveries, such as genetic cloning and development of methods within DNA-based chemistry as well as molecular mechanisms of expression of genetic information were recognized by prizes in Chemistry as further discussed in Chapter 7.

The part of the will that has rarely been possible to fulfill is that the prizes shall be given "… to those who during the preceding year have … ." In practice this requirement has been interpreted to mean that the impact of the contribution to be awarded has been fully appreciated during the preceding year. As a consequence discoveries to be honored generally have been made 5–20 years before the year of awarding the prize and sometimes even much earlier. The problems associated with these delays in recognizing discoveries in Physiology or Medicine is discussed in Chapter 6. The few obvious mistakes that have been made often represent a too quick recommendation for a prize by a committee.

Another part of the will that requires continuous deliberations is that the contribution shall "be beneficial for mankind." The way the committees historically have interpreted this is that high quality basic research leads to discoveries which later, in one way or another, allow applications markedly advancing our civilization. However, the tasks for the committees become more and more challenging. The number of scientists engaged in research increases with time and the number of paradigmatic discoveries probably also increases as a consequence.

The decision on a Nobel Prize cannot be contested nor be the subject of any appeal. The committees and the prize-awarding institutions never respond to complaints.

The awarding institutions and the prizes

The Nobel Prizes represent a major asset of the Swedish scientific establishment. It is a truism to note that the awarding institutions carry a huge responsibility. Only the highest quality in the selection process can be accepted. Thus, the best scientists that can be mobilized in a field in Sweden need to become engaged in the work. Also the selection of international advisors needs to be made wisely. It is not enough for them to highlight a contribution as impressive. The question is, how impressive and how revolutionizing, and furthermore what priority rights should be allotted to the finding by a proposed individual. The need for judges and referees to be well informed and to have good professional networks is a very positive incentive for the society of Swedish scientists. Swedish science also profits in other ways by the national responsibility for prizes in natural sciences. International collaborations are readily established and high-class lectures and conferences can be arranged at Swedish academic institutions, in particular at the prize-awarding institutions, the Royal Swedish Academy of Sciences and the Karolinska Institute.

The financial resources provided yearly by the Nobel Foundation are used for two purposes. It was stipulated in the will that these resources should be divided so that 60 percent are used for the prizes and the rest for the work by the committees and for administration. The work by the committees includes both the formal processing of the nominations received and development of the capacity of the scientists engaged in the work. It is important to secure that they are well informed about the moving frontiers of science.

Money not used a certain year can be reserved for other purposes. Originally there were plans to establish Nobel Institutes which could carry out experiments to control the proposed discoveries. Formally the prize-awarding institutions are obliged to give at least one prize every fifth year. Thus there might have been a temptation for the institutions involved to refrain from giving a prize in a certain year and to put the money back into their own coffers to support their research enterprises. However they only succumbed to such a withholding of prize money on a few occasions and only during the early decades of the selection of prize recipients. In practice, for example, there have been only nine years when no Nobel Prizes in Physiology or Medicine have been awarded. In seven of these years, 1915–1918 and 1940–1942, the reason was the two world wars (see Figure 2, page 148). In the confusion in Europe in the fall of 1914, when the prize was to be announced, it was decided that a temporary delaying of the process would seem to be in order. Eventually a postponement of the prize to 1915 was decreed. In this year Robert Bárány was awarded the 1914 prize in Physiology or Medicine "for his work on the physiology and pathology of the vestibular apparatus." At that time Bárány, an Austro-Hungarian, was a prisoner of war in Siberia. The Swedish Red Cross intervened and managed to get him released so that he could receive his prize through diplomatic channels.

Besides the seven war years without any prizes there were two, 1921 and 1925, when the Institute, during the period of the committee chairmanship of the quality-conscientious Johansson, decided to use the prize money for investments in science at the Institute. A few Nobel institutes were established, but they never came to serve the intended purpose. In time they became obsolete and instead "virtual institutes" have evolved. Thus the extra money is not used for bricks and mortar but instead, for example, for arranging conferences on particularly interesting fields of science where major advances are being made or inviting back previous Nobel Laureates to provide perspectives on developing fields. Needless to say it is not difficult to attract speakers to come to Nobel symposia or conferences!

The Nobel Prizes receive an enormous attention both nationally and internationally. Probably it is the event for which Sweden is best known, with the exception for the temporary blooming of stars in sports. The Nobel Prize is the only prize announced on the front page of dominating daily newspapers and on prime time news on television. Thus it represents a unique opportunity to bring science to schools and to the public, something the prize-awarding institutions attempt to exploit with good judgement. More recently it has become possible to relay the press conferences announcing the prizes and the Nobel lectures in Stockholm in real time on the web. By a requirement in the will Nobel Laureates shall give a lecture within six months after the prize has been received. In practice they are nowadays given two days *before* the prize ceremony on December 10 out of courtesy to the laureates-to-be.

The secrecy of the prize selection process

The selection process for Nobel Prize recipients is surrounded by a highly developed secrecy. This is a prerequisite to endow the process with as high a degree of objectivity and integrity as is possible in human endeavors. Thus, lobbying is useless in affairs that concern Nobel Prizes. If anything, such actions may have a negative effect. After a time lag of 50 years the archives of the Nobel Committees become available for scholarly investigations. These archives contain the nominations, which may vary from a few lines to a thorough review of the field selected for the proposal. The nominations often include valuable supporting documents. They also contain all the reviews initiated by the committees. Some of these reviews are impressively comprehensive and include many critical references. Finally the archives also include the protocols of committee meetings. However these are decision protocols, which generally do not include the views of individual members or the diverging arguments they may have expressed. The protocol of the concluding committee meeting at the Karolinska Institute until 1959, the last year available for archival research, lists the strong candidates and pronounce whether they are prize-worthy. In the concluding paragraph the committee presents the prize recommendation for the year and a brief motivation of the proposed prize. In some cases, as can be seen in Chapters 3, 4 and 6, a variable number of members of the committee may express a difference of opinion from the majority.

At the Academy of Sciences the deliberations and the conclusion by the committee are spelled out already in the book including all relevant documents prepared towards the end of August. It should be noted that with the exception of a number of nominations which may be in English, German or French, all the materials of the archives are in Swedish. This is changing with time since during the last 50 years an increasing number of international reviewers are being used and furthermore also Swedish reviewers may use English — the accepted language of science — upon recommendation by the committee. During the first 50 years the amount of information of the archives shows a clear increase with time, presumably because of the fact that an increasing number of nominations have been considered. Hence it can be predicted that studies of the archives in the future progressively will become even more rewarding.

The number of prize recipients

It was not clear from the will if the prize-awarding institutions should aim at selecting a single recipient or if there could be more than one. Originally the possibility of allowing a split into three prizes per discipline was considered, but eventually it was decided that there could be a maximum of two distinct prizes. It was furthermore not originally regulated whether a single prize could be shared by one, two or more recipients. This was not settled until 1968, when a rule was introduced that there can be a maximum of three prize recipients in one discipline at the same time. This gives five different possibilities. A prize can be given for a single discovery to one person or shared equally between two or three persons. Alternatively a prize can be given for two distinct discoveries. One half of the prize may go to one recipient and the other to another, but one half of the prize may also be shared between two individuals, giving a total of three recipients.

The Nobel Prize cannot be given posthumously. However, if a prize has been awarded in October and the recipient dies soon hereafter he may be honored at the prize ceremony in December. A Nobel Prize does not necessarily have to be given to individuals. It can also, in principle, be given to institutions. In practice this possibility has only been used for the Peace prize, which on several occasions has been given to institutions. One example is the Red Cross, which in fact has received three prizes, one to its founding father, Jean Henri Dunant in 1901 and two more prizes in 1917 and 1945. There are separate rules for the other prize-awarding institutions that prohibit awarding other than individuals.

Table 1 summarizes, in 20-year intervals, the number of recipients of Nobel Prizes in Physics, Chemistry and Physiology or Medicine during 1901–2009. During some time intervals a reduced number of awards were given because of the two world wars or for other reasons. In Physics the first 50 years show predominantly single prize recipients, but there are also a fair number of

Table 1. Number of Nobel Prize recipients.

Time period	Physics			Chemistry			Physiology or Medicine		
No. of recipients	1	2	3	1	2	3	1	2	3
1901–20	15	3	1	16	1	0	14	2	0
1921–40	12	5	0	12	5	0	11	5	1
1941–60	10	6	2	13	4	1	6	6	6
1961–80	6	5	9	11	6	3	2	5	13
1981–00	4	7	9	8	5	7	4	9	7
2001–09	0	2	7	2	1	6	0	4	5
Total	47	28	28	62	22	17	36	32	32
Grand total			187			157			195

shared prizes. After the Second World War the prizes are increasingly shared between two or three recipients. However, there is still today a fair number of prizes with less than three recipients. This is of interest to consider in the light of the increased importance of team work in modern natural sciences. Frequently publications have many hundred authors and one would think that it would be hard to single out individual contributions. This is probably not true. There is always in a team a single person or a few individuals that lead the group.

In chemistry as in physics there is some shift over time towards more than one (all in the first decade) or two recipients after the Second World War. In fact also in this field the distribution of prizes to one, two or three individuals has been rather stable during the last decades. Physiology or medicine is the field in which over time the largest proportion of prizes, some 30 percent, has been given to three individuals. However it seems that the propensity towards selecting three prize recipients may not be increasing with time during later decades. Thus even in a field of multidisciplinary nature like medicine there is, not infrequently, the single or two individuals, who make the difference in paradigmatic advances.

The nationality of Nobel Prize recipients

The total number of personal awards in Physics, Chemistry and Physiology or Medicine until 2009 is 539 (Table 2). However, there is no restriction on a single person receiving repeated honors. Since there are three cases in which a single individual has received two prizes; in Physics to John Bardeen in 1956 and 1972, in Chemistry to Frederick Sanger in 1958 and 1980 and finally in Physics and in Chemistry to Marie Curie in 1903 and 1911, respectively, the true number of individuals who have been awarded a prize is 536. The majority of the awarded scientists have done their work in the USA; 46 percent of the total. In Physics and Physiology or Medicine the figures are close to 50 percent, whereas in Chemistry it is about 40 percent. The figures for prizes in different disciplines given to scientists from Great Britain and Germany are similar, however with a clear dominance for the former country in Physics and in Physiology or Medicine. If the nationality representation is analyzed for separate 25-year intervals (Table 3) it becomes apparent that the good position for Germany predominantly is a pre-World War II phenomenon, whereas since that war the USA has taken an exceptional lead. During the last 50 years

Table 2. Nationality of Nobel Prize recipients.

Country	Physics	Chemistry	Physiology or Medicine	Total
USA	92	62	97	250
Great Britain	23	27	26	76
Germany	18	30	15	63
France	12	8	9	29
Russia (Soviet)	9	1	2	12
Scandinavia	7	8	12	27
Others	26	21	34	82
Grand total	187	157	195	539

Table 3. Natural sciences prize recipients divided into 25-year intervals.

Country	1901–25	1926–50	1951–75	1976–2000	2001–09
USA	4	27	73	106	42
Great Britain	11	17	28	11	8
Germany	24	15	10	10	4
France	10	5	5	5	4

roughly 70 percent of all prizes in natural sciences have gone to this country. Time will tell if the USA can maintain this position.

There have been only a few previous studies[11] of the importance of the nationalities of Nobel Laureates. Other more extensive studies[12] have been concerned with the scientific elite and the sociology of Nobel Prizes. The latter study only dealt with prize recipients from the US. In total 72 laureates were interviewed. Two important phenomena were identified. One was that more than half of all prize recipients had worked in the laboratories of other laureates or laureates-to-be and the other was that there was a pronounced overrepresentation of scientists with a Jewish ethnic background. The current tally is that the representation of Jews is 47 out of 187 laureates in Physics, 25 out of 157 laureates in Chemistry and 49 out of 195 laureates in Physiology or Medicine. Percentage-wise the figures are about 25, 16 and 25 respectively which should be compared to the figures of 2% of Jews in the US and of 0.25% globally. This observation raises the question of how effectively the potential creative capacity of the whole of humanity is used. The fact that all humans, some 2,500 generations back, have a common black ancestor, who moved out of Africa, means that all humans on average have the same potential for creative contributions. If this fact is combined with the observation that a single ethnic group representing only 0.25% is responsible for 25% of prizes in some disciplines one is forced to draw the conclusion that only one percent of all globally accessible creative scientific capacity of mankind is used! A further discussion of the impressive overrepresentation by scientists with a Jewish background and the possible reasons for its somewhat uneven occurrence in different disciplines can be found in the book by Zuckerman[12].

The awardee and the prize

Nobel Prize recipients receive a large sum of money. However the absolute value of the prizes has varied markedly (Figure 3). It had its lowest relative value immediately after the First World War, in 1919, only 28 percent. During the last few decades the prizes have regained the original relative value. A particularly important increase of the volume of the assets managed by the Nobel Foundation occurred towards the end of the 20th century.

There have been continuous amendments of the rules which guide how the Nobel Foundation can manage the assets. Nobel's will states that only investment in safe securities was allowed and this rule was interpreted conservatively. Hence the value of the money decreased, in particular during the two world wars. It was not until 1958 that the original restrictions on investments gradually became liberalized. From then on the Foundation could invest also in stocks, with some limitations, and in real estate. Another factor that restricted the growth of the endowment was that the Foundation had to pay taxes. This was the result of the flaws in Nobel's final will that forced the creation of the Nobel Foundation as the recipient of the money for the prizes. This situation lasted until 1946 when the Foundation eventually successfully negotiated an exemption from all Swedish taxes except local real estate tax.

In the late 1980s the Nobel Foundation managed to markedly increase the size of its endowment. There is a particular twist to this story. The Foundation had invested a considerable portion of its assets in buildings in Stockholm. However, there was a problem with this investment, since, as was mentioned, local real estate taxation was not exempted. The Executive Director of the Foundation, the late Stig Ramel, had a discussion with the Finance Minister at the time, Kjell-Olof Feldt, arguing that the management of the Nobel assets

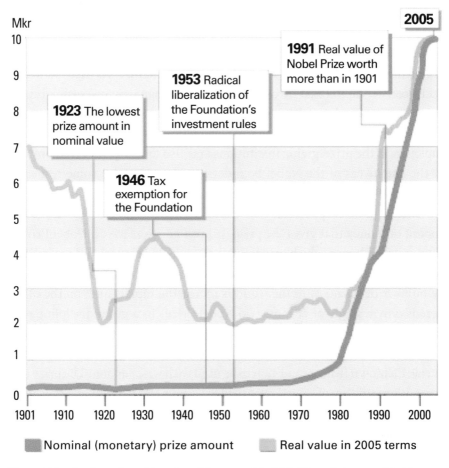

Diagram comparing the nominal (monetary) prize amount
since 1901 with its real value in 2005 terms — in SEK per unshared prize

Mkr
2005

1991 Real value of
Nobel Prize worth
more than in 1901

1953 Radical
liberalization of
the Foundation's
investment rules

1923 The lowest
prize amount in
nominal value

1946 Tax
exemption for
the Foundation

Nominal (monetary) prize amount   Real value in 2005 terms

Figure 3. The absolute value of the prizes. [Courtesy of the Nobel Foundation, modified.]

represented an exceptional case because of the international prestige of the prizes. Ramel proposed that all the capital gains should be tax exempted. When the Minister answered that regrettably this was not possible, Ramel asked for an advice. All the Minister could answer was "Sell it all." So Ramel did this, and three months later the real estate market collapsed; so much for the science of economy and investments. The Foundation managed to triple its assets and in 1991, 90 years after the first prizes were given, the value of the prize was returned to its original relative value. Today it even surpasses this value. 1991 was the last year in which Ramel was leading the Foundation and coincidentally it decided to have a special celebration and invited previous Nobel Laureates to the festivitas. With the particular mathematics of the Foundation this 91st year of prize awarding was referred to as a 90th year celebration, since that was the time that had passed since the first prizes were awarded.

During the preceding year the Board of the Foundation decides the value of the prizes for a certain year in Swedish crowns. The value in 2009 was ten million Swedish crowns, well in excess of one million US dollars. The yearly yield from the capital that the Nobel Foundation manages to generate is used not only for the prizes (close to 60% of the returns should be used for this purpose) but also to pay for work performed by the committees and other employees at the prize-giving institutions and also for the prize ceremony and for the employees of the Nobel Foundation. There are also some resources made available for the Nobel institutes, as mentioned above.

It should be emphasized that to the awardees it is not the handsome amount of money that gives the prize its prestige. To receive a Nobel Prize is an unmatched honor, which brings an unprecedented respect and recognition from colleagues. This is further accentuated by the fact that on the one hand the number of scientists in the world is increasing rapidly and, on the other, there is only a single or at most two Nobel Prizes in a given discipline each year. In fact Nobel Laureates, in addition to recognition by the scientific establishment, acquire a prestige to make authoritative statements also in matters outside their own field, a condition to be used with wise caution. The prize also serves as recognition of a particular field of science and may enhance advances in this field.

Of course the press frequently asks Nobel Laureates trivial question about what they are going to do with the money. In many cases the money is used to stimulate young scientists by the establishment of a foundation or by some other means. Werner Arber, who shared a prize with Daniel Nathans and Hamilton Smith in 1978 "for the discovery of restriction enzymes and their application to problems of molecular genetics" used the money to invite a large group of his friends to walk the Swiss Alps and discuss science and other essential aspects of life. My favorite story when it comes to the use of the prize money involves Günter Blobel. He was the single recipient of the prize in Physiology or Medicine in 1999 "for the discovery that proteins have intrinsic signals that govern their transport and localization in the cell." He had personally experienced the devastative destruction of German cities during the Second World War and therefore decided to use the money to assist in the rebuilding of the Frauenkirche in Dresden. This church now stands in its original splendor. However, not having any religious motives for his engagement Blobel decided to give equal credit to the cultural contributions of all the three monotheistic religions. He consequently also invested money in the building of a synagogue and a mosque in the same city.

The time lag between discovery and prize awarding means that the recipients are frequently well advanced in their career (see Chapter 6). Obviously the money does not serve the intended purpose of providing financial independence to further continued long time research. Still, there are some Nobel Laureates who keep on being highly productive in their science after they received the prize. Paradoxically there is a drawback for a scientist to receive the prize when he (or she) is too young. He will not receive any further prizes, since no prize can compete with a Nobel Prize! I learnt this when I hosted one of the Nobel revisiting lecturers at the Karolinska Institute. The committee at that Institute invites back previous laureates to get an update of the advance of science and to discuss various potential fields for awarding.

The age of Nobel Laureates at the time when they receive the prize in Physiology or Medicine varies between 32 and 87 years (Figure 4). During the first 50 prize years (1901–1959) the average age was 53.9 years (76 laureates), but during the following 50 prize years it was higher, 59.1 years (119 laureates).

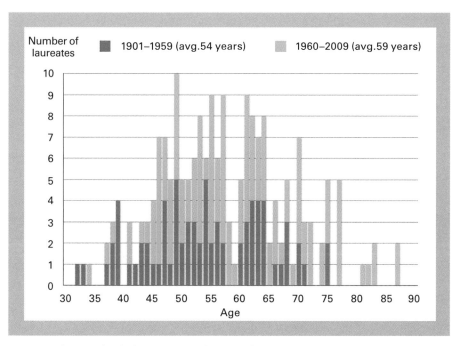

Figure 4. The age of Nobel Laureates at the time when they received the prize in Physiology or Medicine.

The three youngest prize recipients are Frederick G. Banting (1923), 32, and Joshua Lederberg (1958), 33 (both discussed in Chapter 6) and James D. Watson (1962), 34 (Chapter 7). One of the oldest laureates until 1959 was Charles S. Sherrington. He had to wait a long time until he eventually, at the age of 75, received his prize in 1932 (also mentioned in Chapter 6). Since 1960 prize recipients in a number of cases have been over 80 years of age. The record age, held by two laureates, is 87. One of these is Peyton Rous, discussed briefly in Chapter 3. Thus most prize recipients in Physiology or Medicine are of mature age when they become laureates and their average age has increased during the previous century. Possibly, there are less controversial explanations for this phenomenon, such as the progressively increased life span and period of creativity during the preceding century plus perhaps the increased complexity of the knowledge field to be covered before it is possible to make a major contribution. Another concern that has been ventured is that the prize has a paralysing effect on creativity. I do not believe that this is a large problem. Many of the prize recipients continue to make their mark on the field they have pioneered and assumingly most of them use their elevation to a unique rank of statesmen of science in different positive ways.

Finally, there is of course also a unique social dimension to becoming a laureate. The 7–10 days in Stockholm in dark December — the prize ceremony is always on Nobel's death day, the 10th of the month — is a fairy tale experience. It cannot be described, but only experienced. In the 100-year jubilee in 2001 (meaning 100 years *after* the first prizes were awarded) all previous laureates were invited to come to Stockholm (see this book cover). One wonders if ever previously in the development of human civilizations had so much intellect been gathered in one and the same place.

There is no Nobel Prize in economics

At the end of the 1960s the Swedish Central Bank had a tercentennial celebration. It then offered to donate an annual sum of money to the Nobel Foundation in order to allow the Royal Swedish Academy of Sciences to give a prize in economic sciences in memory of Alfred Nobel. This Academy has a class for economic and social sciences which could manage such a prize. The long experience the Academy had in managing the Nobel Prize in Physics and Chemistry was seen as a valuable asset in this context. This proposal was accepted by the Academy, by the Nobel Foundation and by relatives of Alfred

Nobel who were consulted, and approved by the Crown. *The Sveriges Riksbank Prize in Economic Sciences in Memory of Alfred Nobel* has been awarded at the same ceremony as the one for the true Nobel Prizes since 1969. Since it is impossible for the public or the press to use this long name and since the prize even in the minute details is treated as one of the true prizes specified in Nobel's will it is generally referred to as the "Nobel" Prize in economics.

In the statutes for the prize it is specified that "it shall be awarded annually to a person who has written a work on economic sciences of the eminent significance expressed in the will of Alfred Nobel drawn up on November 27, 1895" and further "The prize shall be awarded by the Royal Swedish Academy of Sciences in accordance with the rules governing the award of the Nobel Prizes instituted through his will." Since apparently the terms "discovery," "invention" or "improvement" used by Nobel in his will for the prizes in natural sciences do not apply, it remains to define the meaning of the term "eminent significance" that is used. Maybe the closest analogy is with "outstanding work" used for specification of the prize in Literature. The latter prize, however, represents a very special form of evaluation of the depth and width of human behavior cultures.

The prize for economic sciences is treated like the true Nobel Prizes on the home page of the Nobel Foundation, Nobelprize.org and also in the annual book *Les Prix Nobel*, where the introductory speeches at the prize ceremony of the recipients of the Prize in Economic Sciences, the autobiographies of the laureates and their official lectures are printed. An example of indistinguishable treatment of these different prizes is the awarding of a jetton to those who participate in the decision of a prize. The awarding of such a jetton is specified in Nobel's will and originally it was made of solid gold. In the 1970s it was discovered by the meticulous Swedish tax authorities that the academicians who participated in decisions about Nobel Prizes, in addition to their already modest income, could receive an annual extra remuneration. It was therefore decided that the small medallions no longer should be of solid gold but be gilded. They were referred to as "Ramel's emergency money," after the above-mentioned Stig Ramel. The same kind of "emergency money" is handed out to members of the Academy who participate in the decision on a recipient(s) of the Prize in Economic Sciences. Ten of these gilt jettons, disregarding for which prize decision they are given for, can be exchanged for one of solid gold.

The yearly decision that the board of the Nobel Foundation makes about the value of the prize money for the forthcoming year has important consequences for Sveriges Riksbank. This bank provides the money for its

Prize in Economic Sciences in memory of Alfred Nobel. According to the agreement the bank should pay the Nobel Foundation the prize sum that the board has determined plus an extra 65 percent to cover the costs for the engagement of the Royal Swedish Academy of Sciences for its work to select the prize recipient(s) and also the management at the Foundation. However it should be noted that the bank has *no* influence on the amount of money to be paid. The bank is not represented on the board of the Foundation. Hence, in 1991, when the value of the Nobel Prizes was returned to its original value, the bank had to pay the requested heavily increased sum of money. It is not likely that this extra payment made the bank insolvent, but still one would like to hope that this agreement is exceptional. Finally it should be noted that Sveriges Riksbank does not provide any extra money corresponding to that for the "virtual Nobel institutes" that is available for competence upgrading of the scientists involved in the work with the true Nobel Prizes.

About a decade after the establishment of the Prize in Economic Sciences in memory of Alfred Nobel, the Royal Swedish Academy of Sciences was approached by Holger Crafoord, a successful industrialist and philan-thropist who wanted to donate money to large prizes in fields not covered by the Nobel Prizes. This was accepted by the Academy and since more than 25 years ago the Academy has awarded prizes to celebrate major advances in the disciplines of mathematics, astronomy, geology, biology, with special emphasis on ecology and also, because of the donator's own experience of this disease, the special field of medicine concerning polyarthritis. These prizes, each worth 500,000 US dollars, are awarded at alternating years in a ceremony that is also honored by the presence of the Swedish King. One may wonder if, in a historical perspective, a strong and independent discipline like economic sciences would not have fared better by not borrowing the luster and reputation of the Nobel Prizes and instead been treated like the Crafoord prizes.

The enigmatic concept of discovery

As will become apparent by the following chapters the key definition on the quality of research in natural sciences to be recognized by Nobel Prizes is discovery. A discovery comes unexpectedly and is unanticipated. The lessons it sends are that one cannot plan for major breakthroughs in science. It concerns a phenomenon of major importance and its unraveling has a huge impact and revolutionizing consequences for the way science is conducted after it has been made. It leads to the opening of a new field of science and there is a rapid acceleration of publications in this new field. Many have speculated about the conditions concerning both individuals and institutions that may foster creativity and excellence. In several of the chapters I will return, directly or indirectly, to the question about how a discovery is defined and to the conditions that may further its emergence.

Creativity in both the sciences and the arts is an enigmatic phenomenon. Perhaps it is inherent in its nature that it cannot be induced, projected or conjectured. Still we want to understand how we can stimulate creativity and an endless number of books have been published on this theme. Since 2001 a Nobel Museum has been established in Stockholm. This museum is not under the aegis of the Nobel Foundation, since by the legal rules of this organization it must be involved exclusively with providing support to the prize-giving institutions. Thus the Nobel Museum and the home page of the Foundation, nobelprize.org, are managed and financed by a separate legal entity. However the overlap of representatives on the boards of the Foundation and of the organizations managing the latter operations secures a harmonization of efforts.

The first exhibition in the Nobel Museum had the theme "Cultures of Creativity: Individuals and Milieus." The main parts of the exhibition were the following: individual creativity, creative milieus, the Nobel system, Alfred Nobel and his time, and the Nobel Laureates 1901–2001 (samples only). In addition to the exhibition in Stockholm an identical version of it has been touring the world for display in large cities in Europe, USA and Asia. The exhibition tried to catch the enigma of creativity, but it probably raised more questions than it gave answers to. The charm of human endeavors that change our civilization is that they are unpredictable.

It is again a truism to note that it is unique individuals who contribute to the advance of science. The exodus of scientists from Europe to the US, in particular Jews from Germany and other European countries, in connection with the Second World War, was an important factor in the development of the hegemony in science of the latter country. But there are also other factors. Resources help, but more important is an intellectual density created by aggregation of minds and a complete freedom of these minds to exchange ideas. It is striking that certain institutions have spawned a large number of Nobel Laureates[13,14], such as the Rockefeller University (former Institute), the California Institute of Technology and the Laboratory of Molecular

Biology in Cambridge, UK. All these institutions are characterized by a loose organizational structure with emphasis on research groups and their leaders. The intellectual and personal leadership is extremely important. A large percentage of Nobel Laureates have worked in laboratories of other laureates.

Sune Bergström (1916–2004), recipient of the Nobel Prize in Physiology or Medicine 1982. [From *Les Prix Nobel 1982*.]

Another important factor is what might be called the democratization of science. This phenomenon evolved in particular in the USA after the Second World War and contributed to the dominance of this country in the natural sciences. Professor Sune Bergström, Vice-Chancellor of the Karolinska Institute (1967–1977), Chairman of the Board of the Nobel Foundation (1975–1987) and himself a Nobel Laureate in Physiology or Medicine in 1982, commented on this in his opening address at the prize ceremony in 1976[15]. He stated:

> But apart from these basic economic prerequisites, there are other actors which have greatly contributed towards the rapidity of America's expansion in the research sector and which are perhaps of particular interest to Europeans. Universities all over the world expanded rapidly during the 1940s and 1950s. In many places, not the least the European countries, this expansion took place with the retention of traditional hierarchical and inflexible structures.
>
> In the USA, on the other hand, the growth of university research was characterized by a dynamic openness in forms which might be characterized as a democracy of research workers. Many visitors to American institutions and scientific congresses after the war were struck by the natural way in which professors and students could conduct scientific discussions on a basis of equality and also by the practice of making young researchers responsible at an early stage in their careers for independent research projects with the big institutions.

To summarize, in milieus of a non-dogmatic nature and with non-authoritarian characteristics unexpected things can happen if the right minds are present. They have to be prepared to grasp the unexpected opportunity. Serendipitous events have often changed the course of science, as is discussed in the next chapter.

The aging Alfred Nobel. [Courtesy of the Nobel Foundation.]

Coda — Alfred Nobel's surprise

What would Nobel have thought if he had known about the global interest in his prizes? The answer to this question is probably that he would have been very surprised. His intention to provide a long-lasting scholarship for talented young scientists has not been fulfilled. Instead the prizes have generally been bestowed on mature scientists for their fundamental discoveries. The awards have evolved to acquire global visibility, to be the gold standard of all prizes. They have truly marked the amazing advances in the sciences during the 20th century and can be expected to continue to do so in the 21st. Futhermore they have made science visible to society at large. Their role may take on increasing importance as we move into societies with an ever-increasing dependence on the advance of science. We can take advantage of the Nobel Prizes as a means both to reveal to the public at large that the lifestyle of scientists can be uniquely rewarding — they can learn amazing things — and that there are often unexpected applications of new knowledge which can change the way we lead our lives. To this should be added the fact that new unraveled knowledge helps us also to reflect on our personal existential questions.

The Royal Swedish Academy of Sciences has commemorated Alfred Nobel with two medals, one as early as 1902 but the other in 1996, one hundred years after his death. The inscription on the first medal reads NATURA INVENTAS VITAM IUVAT EXCOLUISSE PER ARTES (it is a delight to ennoble human life by art and inventions) and the one on the second medal simply reads CREAVIT ET PROMOVIT (he created and promoted). This retrospective attention is certainly well deserved. But the scientific community has also honored Nobel by naming one of the unstable elements after him, Nobelium. It is very likely that Alfred Nobel would have enjoyed knowing that the prizes specified in his will have had such a remarkable impact, so far, for more than one hundred years. In a rough parallel with the adventures of the Princes of Serendip, central figures in the next chapter, it can be said that Nobel's goal of establishing a powerful subsistence support for promising young scientists never materialized, but instead he came to be the father of the world's most famous prizes for the sciences, literature and peace.

Chapter 2

Serendipity and Nobel Prizes

Nitroglycerin is a liquid explosive discovered in 1847 by Ascanio Sobrero, an Italian chemist. People were not slow to realize its many potential applications, such as blasting rocks away during road construction, or burrowing tunnels through mountains in the development of railway systems. However, harnessing the explosive power of the compound was a major problem, and research focused on its solution. One day, a scientist working in the laboratory happened to cut his finger on a piece of glass. As was common practice at the time, the wound was covered with collodion, by exposing the finger to cellulose nitrate dissolved in ether and alcohol, and then letting the solvents evaporate. That coating was quite flammable, and in another, more highly nitrated form, cellulose nitrate was in fact used as an explosive. It then struck the scientist that nitroglycerin and nitrocellulose might be combined into an explosive much more powerful than dynamite — a mixture of nitroglycerin and kieselguhr — but still just as stable. Early the next morning he began to experiment with different proportions of the two compounds in the laboratory in his home in Paris (page 40). When his assistant arrived for the day's work, our scientist showed him a jelly-like mixture of the two most powerful explosives known at the time.

The scientist was Alfred Nobel, and the new explosive formula was developed into blasting gelatine, patented in 1875 in England, and a year later in the US. This accidental discovery contributed markedly to the success of Nobel's business in explosives, which made him a wealthy man. In his last will

Nobel's house in Paris on Avenue Malakoff. It was used both as his home and laboratory (1873–1881). [Courtesy of the Nobel Foundation.]

he bequeathed his large fortune to allow annual awards of international prizes in five different fields. The rest — as the saying goes — is history. But Alfred Nobel's discovery of the blasting gel brings us to the central theme of this chapter, the phenomenon of serendipity. His discovery was a "happy" accident, immediately appreciated by its inventor as being of potential value.

Soon after its discovery it was recognized that nitroglycerin had other properties, in addition to being an explosive. It shows vascular activity and therefore came to be used for relaxing contracted blood vessels in the heart and as already mentioned it was recommended for use by Nobel himself when he developed his heart problems. It would take almost another 100 years before the mechanism of the action of nitroglycerin on blood vessels was explained. In 1998, a Nobel Prize in Physiology or Medicine was awarded to the US scientists Robert Furchgott, Louis Ignarro and Ferid Murad "for their discoveries concerning nitric oxide as a signaling molecule in the cardiovascular system." In experiments with a surprising outcome, these researchers had managed to demonstrate the mechanism of action of nitroglycerin. The compound acts on the carpet of cells lining the inside of blood vessels, the endothelial cells, causing the release of the active component, which turned out to be a gas, nitric oxide. This finding was quite unexpected. The discovery led the pharmaceutical industry to initiate a search for related molecules that could also cause the release of nitric oxide. The most active compounds were then tested for their capacity to relieve angina pectoris, in patients with heart disease. In one of these trials, a compound was identified that not only affected the blood vessels of the heart but also those of the penis. It furthered erection — and it did not take long before Viagra had been developed. Today this is one of the most extensively used drugs in the world.

From left: Robert Furchgott, Louis Ignarro and Ferid Murad, recipients of the Nobel Prize in Physiology or Medicine 1998. [From *Les Prix Nobel 1998.*]

In this chapter the role of accidental discoveries in the advance of science will be discussed, but first some comments on the term "serendipity," the man who invented the word, and its applications to scientific endeavors through time.

The origin of "serendipity"

The English aristocrat and man of letters Horace Walpole corresponded for 46 years with his distant cousin Horace Mann, who spent most of his life abroad in the British diplomatic service. In a letter of January 28, 1754, Walpole wrote apropos a recent "critical discovery":

> This discovery I made by a talisman, which Mr. Chute calls *sortes Walpolianae,* by which I find everything I want, *a pointe nomminee* (at the very moment), wherever I dip for it. This discovery, indeed, is almost of the kind I call *Serendipity,* a very expressive word, which, as I have nothing better to tell you I shall endeavour to explain to you: you will understand it better by its derivation than by definition. I once read a silly fairy tale, called *the three Princes of Serendip;* as their Highnesses travelled, they were always making discoveries, by accidents and sagacity, of things which they were not in quest of: for instance, one of them discovered that a mule blind of the right eye had travelled the same road lately, because the grass was eaten only on the left side, where it was worse than on the right — now do you understand *Serendipity?* One of

the most remarkable instances of this *accidental sagacity* (for you must observe that *no* discovery of a thing you *are* looking for comes under this description) was … .

The "silly fairy tale" that Walpole alluded to first appeared in 16th-century Venice under the title *Peregrinaggio di tre giovani, figliuoli di Serendippo, tradotto dalla lingua persiana in lingua italiana de M. Christoforo Armeno* (translated into English in 1722 as *Travels and Adventures of Three Princes of Serendip*). The three princes were sons of Jafer, the philosopher-king of Serendip, a former name of Ceylon, present-day Sri Lanka. He wanted them to travel and learn about the customs of other people as a complement to their book knowledge. In contrast to what is frequently cited, they were not in search for treasures, and during their travels they experienced adventures and made various discoveries. Many of these were made by use of their keen observations, which often resembled Sherlock Holmesian insights. One of them concerned a camel — in the original story it was not a mule as Walpole remembered. Without having seen the animal, the princes would deduce not only that it was one-eyed, but also that it lacked one tooth, that it was lame, that it was laden with butter on one side and honey on the other, that it was ridden by a woman, and that she was pregnant. As an example: the load of the animal was deduced from the observation that there was a trail of ants, which love butter, on one side of the road, and of flies, which treasure honey, on its other side. As emphasized by Walpole, the discoveries always concerned things for which the princes were not looking. As will become apparent the practical application of the term to scientific discovery is often wider and includes both phenomena not searched for and those searched for.

Horace Walpole

Horace Walpole lived from 1717–1797, was of aristocratic descent and became the 4th Earl of Orford. He was a man of independent means — like some of his contemporary gentlemen scientists — who could satisfy his needs for a comfortable life, and filling his estate, Strawberry Hill, with the strangest things. Thus he was an indiscriminate rather than eclectic collector of trivia. Apparently he was also a collector of words, and to satisfy this urge he was an inveterate inventor of new terms. He obviously had a penchant for neologisms. A later biographer described him as a downright trifler, too delicately constituted

Horace Walpole (1717–1797). [Engraving by James McArdell after the portrait by Joshua Reynolds (1757).]

for real work in literature and politics. Still he served for 27 years as a Whig member of the Lower House of Parliament. In addition, this representative of the idle class was a prolific writer and correspondent. He published a number of books, often on current political issues, which he had printed in his own press. His style was characterized by acumen and epigrammatic wittiness, but it was also pretentious and showed a propensity for paradoxical formulations. His vast collection of letters was published in several volumes between 1818 and 1833, those written to Horace Mann appearing in the final volume. So it was in 1833 that the word "serendipity" came to light for the second time, but only to again take on a Sleeping Beauty existence.

Serendipity and scientific endeavor

In the early 18th century there was a considerable interest in both England and France in Oriental history and literature, and themes from the Orient were widely used — e.g. by the Enlightenment philosopher Voltaire. In a novel, the fame of which is perhaps surpassed only by *Candide,* he wrote about a philosopher in ancient Babylonia, *Zadig* (which is also its title). The book challenges religious and metaphysical orthodoxy based on the moral revolution that takes place in Zadig, when he reflected on the facts of destiny beyond human control. In parenthesis it can be mentioned that Zadig & Voltaire is the name of a French chain of luxury boutiques for clothes and utilities, represented also in Stockholm. In 1880, Thomas H. Huxley, the pre-eminent proponent of Darwin's new theories of evolution by natural selection, wrote an article "The method of Zadig," which contains the quote: "What, in fact, lay at the foundation of all Zadig's arguments, but the coarse, commonplace assumption, upon which every act of our daily lives is based, that we may conclude from an effect to the pre-existence of a cause competent to produce that effect?" Depending upon the situation of the observer, one and the same discovery may be described — in retrospect — as a prophecy, an event of serendipity, or simply as a happy accident.

At that time the term "serendipity" had not yet been assimilated by scientists; this would take another 50 years. Before that, however, the word had been resurrected by writers. In the English periodical *Notes and Queries,* its meaning was discussed by amateur scholars Edward Solly and Andrew Lang (1878). At the turn of the century its use was somewhat widened, to extend beyond a small circle of literati. This was due to initiatives by a magazine editor,

Wilfrid Meynell, whose son opened the *Serendipity Bookstore* in London's East End in 1929. The word officially appeared for the first time in the unabridged Oxford English Dictionary in 1912.

In the 1930s, "serendipity" eventually took the big leap from the world of letters to the world of science. The central figure in this event was a humanist and medical scholar, Walter B. Cannon, professor of physiology at the Harvard Medical School. Cannon used the word frequently, not only in referring to the phenomenon of accidental discoveries in science but also in discussing the philosophy of scientific pursuits. In his book *The Way of an Investigator*[1], he devoted a full chapter to the word. Also Hendrik van Loon[2] has a section subtitled "A short chapter which for the greater part is devoted to an explanation of the word serendipity" in his book *The Arts*. It focuses on the work of Heinrich Schliemann, the first archaeologist to get a view of Tutankhamen's grave chamber. On that occasion he is reported to have exclaimed "... I see wonderful things!" — probably as (questionably) authentic as Archimedes' *eureka*!

Independently from Dr Cannon, other natural scientists also used "serendipity." Thus Dr Ellice McDonald, director of the Franklin Institute for Cancer Research in Philadelphia, wrote: "While on holiday in Cuba in 1938, I found this word (serendipity) in a detective story by S. S. van Dine and quoted it in my annual report. It seems so pertinent to the vagaries of research that many have used the word since." van Dine was a popular detective story writer in the 1930s, the *nom de plume* for Willard Huntington Wright.

A third example of the emerging use of the serendipity concept in the 1930s is by the eminent physicist and chemist Irving Langmuir, an employee of the General Electric Corporation between 1909 and 1950. He made fundamental contributions to the development of the gas-filled light bulb and the electron tube, and in 1932 received the Nobel Prize in Chemistry "for his discoveries and investigations in surface chemistry." He became a foreign member of the class for technology at the Royal Swedish Academy of Sciences in 1937, where he succeeded the recipient of the 1909 Physics prize, Guglielmo Marconi. Langmuir's leadership of the large General Electric research laboratory was characterized by the principle of openness, which he had adopted from his

Irving Langmuir (1881–1957), recipient of the Nobel Prize in Chemistry 1932. [From *Les Prix Nobel 1932*.]

predecessor Willis Whitney. This is how his leadership concept is cited in the Standard Oil trade journal *The Lamp,* published in 1953:

> Cultivating serendipity is, essentially, a matter of being constantly on the lookout for the chance reaction that may lead to a discovery. Dr Irving Langmuir ... deliberately nurtures serendipity by never setting himself a specific goal. As he puts it, he just 'has fun in the laboratory' ... 'Discovery cannot be planned', Langmuir said recently 'but we can plan work that will lead to discoveries'. In other words, research directors can help create an atmosphere in which the muse of serendipity is most likely to be wooed and won. They often do this by planning programs broad enough to allow their researchers the freedom to follow leads they chance upon.

Robert Merton (1910–2003), Columbia University, New York. [Photo from Harriet Zuckerman.]

In the mid 1940s, "serendipity" was introduced into the social sciences by one of its eminent figures, Robert K. Merton. He was a bibliophile who loved to browse at random in the Oxford English Dictionary that he had acquired as a young student. Once, when he was looking, as he jokingly described it, for "sequestration" or "sesquipedalism" (I trust not many readers have encountered the latter word before — it means "1½ foot-length," used metaphorically to designate an individual given to the use of long words!) he encountered the word "serendipity." He then started to use it in his scientific publications, which to a large extent dealt with the sociology of science. He is probably best known for his book *On the Shoulders of Giants,* originally published in 1965[3]. The title is taken from Newton's famous dictum "If I have seen farther, it is by standing on the shoulders of giants." In a footnote, Merton refers to the manuscript of a book *The Travels and Adventures of Serendipity,* co-authored with Elinor Barber and finished in 1958. It rested in Merton's files until it was published posthumously in 2004[4], with a rich Afterword written by him at the age of 91 years. At the time of writing, there were 135 printed sources of the use of the word; the Afterword outlines the explosive extension of its use after 1958. The book by Merton and Barber provided an invaluable source for different discussions in this chapter, including the above-mentioned relations of serendipity and Voltaire's *Zadig.*

In the following are presented selected cases of scientific discoveries, many of which have been awarded Nobel Prizes in Physics, Chemistry and Physiology or Medicine, that exemplify serendipitous events; additional examples can be found in the book by Royston M. Roberts[5].

Serendipitous events in science

The accumulation of new knowledge in human civilizations has passed through different phases. Originally we had to use our five senses to register phenomena in our surrounding milieu to find out what correlations we could identify and if we could distinguish some pattern in our observations. Much of the accumulated information was essential to our daily existence. Which plants were edible, which of them had useful pharmacological effects and which belonged to none of these categories? The botanist Carl Linnaeus, one of the founders of the Royal Swedish Academy of Sciences in 1739, was sent out by the Swedish House of Parliament to make an inventory of plants in various regions of the country, according to these three categories. In the same way humankind wanted to find out which animals could be hunted for food, harnessed for domestic use, and which represented a threat to our existence. But we also wondered about the nature of the life-giving sun, the thunder, the fire or the starry sky. When factual knowledge was lacking, myths arose. These have progressively been exchanged for facts, not the least due to the highly successful hypothesis-driven experimental science that developed during the 18th century. Empirical and often myth-fulfilling science evolved into a new kind of science during the time of the Enlightenment. Thus happy fortuitous observations and accidents played a major role in the dawn of knowledge accumulation. As civilization developed more rational science, one might have expected that the search for new knowledge would become more structured, more predictable and hence less dependent on chance events. As will be presented in this chapter, this is only partly true, and serendipitous events continue to play a major role in scientific discoveries and will continue to do so in the future. Creativity in the arts and sciences shows many similarities. In both domains it depends on the inspiration sparked by accidental events and appreciated by deeply committed, often obsessed individuals.

In addition to the happy accident, the presence of a "sagacious" observer is a necessity — as formulated by Louis Pasteur in his famous dictum *Dans les champs de l'observation, le hazard ne favorise que les esprits preparés* (In the field of observation, chance favors only the prepared mind). Serendipitous and *eureka*-inspiring events, like Archimedes' bath tub, Newton's falling apple and Watt's boiling kettle are readily quoted — though they may be myths invented later by a reverent public. However, in all these cases it was the prepared mind of the scientist/inventor that turned a trivial observation, made by many before, into a lever to a new understanding. The generation and assimilation of knowledge that cannot be identified and quantified by the use of our senses poses particular problems. Here conceptualization is required, as when Newton came to appreciate the force of gravity. In order to manage the necessary abstraction we need to invent useful metaphors. The field of physics in general is a challenge since it is often defined by quantifiable forces out of grasp by our senses.

The magic of invisible rays

Conrad Roentgen (1845–1923), recipient of the Nobel Prize in Physics 1901. [From *Les Prix Nobel 1901*.]

A case in point is the first Nobel Prize in physics, awarded in 1901 to Wilhelm Conrad Roentgen for his discovery of X-rays. Roentgen was interested in detecting cathode rays coming from an evacuated all-glass tube without the commonly used aluminium window. Instead, he covered the tube with black cardboard and used a separate phosphorescent barium platinocyanide screen to detect the rays. He tested the cover, and it indeed seemed to function well in the completely dark room. On his way to turn on the light, he noted a weak fluorescence on the screen that was at almost half a meter's distance from the tube. He knew that this could not have been caused by cathodic rays, since these travel only a short distance, so the fluorescence must have been caused by some new sort of rays! Since he did not understand their nature he called them X-rays. In his preliminary publication of 1895 "A New Kind of Rays, a Preliminary Communication," he wrote: "If the hand be held between the discharge tube and the screen, the darker shadow of the bones is seen within the dark shadow image of the hand itself." Within a year, his discovery had spread for medical use all around the world. It was a lucky choice by the Royal Swedish Academy of Sciences to recognize this discovery by the first Nobel Prize in Physics.

Henri Becquerel (1865–1943), recipient of one half of the Nobel Prize in Physics 1903. [From *Les Prix Nobel 1903*.]

Henri Becquerel was inspired by Roentgen's discovery and reasoned — incorrectly — that certain substances made phosphorescent by visible light might emit other forms of penetrating radiation. To study this assumption, he selected uranium, which he put in contact with a photographic plate wrapped in black paper. He then exposed the crystal to bright sunlight, and an image of the crystal was formed on the film. Due to the fact that the sun did not shine in Paris for a number of days, he had to postpone further experimentation and put the uranium crystal and the covered film away in a drawer. Several days later he took out the materials and for some reason decided to develop the film. To his surprise, it showed a perfect image of the crystal, as if it had been exposed to sunlight. By this accidental observation natural radioactivity was discovered. In 1903, Becquerel received half a Nobel Prize in Physics "for his discovery of spontaneous radioactivity," with the other half awarded to Marie and Pierre Curie.

Patterns of background radiation

The field of astronomy offers additional examples of serendipitous discoveries. In 1964, Arno Penzias and Robert Wilson at the Bell Laboratories used radio antennae to communicate with the early satellites. In this work it was important to eliminate all terrestrial sources of background radio signals. They carefully cleaned their antennae, which included removal of some pigeons and their droppings, euphemistically referred to as a "white dielectric substance." In spite of these precautions some "noise" remained. At about the same time, James Peebles at Princeton University presented a paper, which elaborated on the Big Bang theory for the origin of the universe. This explosion of extremely condensed matter could be expected to have released a huge amount of radiation energy, which should be recognizable in a continuously expanding universe. Connecting the two observations eventually led to the appreciation that the "noise" picked up by the Bell Laboratories antennae did not come from Earth but from outer space. It had just the amount of energy

Arno Penzias and Robert Wilson, recipients of one half of the
Nobel Prize in Physics 1978. [From *Les Prix Nobel 1978.*]

that could be calculated to remain from the radiation disseminated by the Big
Bang. This new opportunity to study the birth of the universe was awarded
with half a Nobel Prize in Physics in 1978 to Penzias and Wilson "for their
discovery of cosmic microwave background radiation."

Jocelyn Bell and Anthony Hewish. Hewish received one half of
the Nobel Prize in Physics 1974. [Courtesy of the University of
Bath and from *Les Prix Nobel 1974.*]

Similarly, Jocelyn Bell and Anthony Hewish at Cambridge University
found something unexpected when measuring radio signals from the universe. Hewish and his colleagues had developed an instrument capable of
registering extremely rapid responses. Originally, the idea was to analyze
effects of the outer solar corona on the incoming signals. To their surprise,
Bell detected a previously unknown pattern of radio signals: short pulses,

repeated periodically with a high degree of precision. This was a new kind of "twinkling" phenomenon originating from an unknown source, which was eventually interpreted as a neutron star with a strong magnetic field. As it rotates, the star emits beams of radiation into the universe, just as a lighthouse emits its rotating beam. This kind of neutron star became known as a pulsar, and the discovery of pulsars was to shed new light on the structure of the universe. The radio waves have been traveling for thousands of million years at the speed of light to reach our Earth.

This serendipitous discovery led to the Nobel Prize in Physics being awarded in 1974 to Hewish, a prize which has been the subject of contentious discussions. The prize did not include Bell, who was a graduate student at the time — instead, Hewish shared it with Martin Ryle. The motivation for their prize was "for their pioneering research in astrophysics: Ryle for his observations and inventions, in particular of the aperture synthesis technique, and Hewish for his decisive role in the discovery of pulsars." Hewish refers to Bell five times in his Nobel lecture, and there is no doubt about the fact that her diligent recordings were critical to the discovery. Hewish clearly played a major role in the development of the radio astronomy instrumentation that allowed the measurements to be made. With his deep insight into the field, he also put the discovery into context, which led to a new understanding of the universe.

In each assignment it behoves a Nobel Committee to decide on the limitations of the discovery under scrutiny (discussed in Chapters 3 and 6). The critical questions are when and by whom the first critical observation was made and further, at which stage in the developments the scientific community has come to a general acceptance of the validity of the discovery. Allocating fair credits to mentor and apprentice, possibly in particular if the latter is a woman, may be difficult. Husband and wife, as closely collaborating scientists, may represent additional problems as regards the judging. Joseph Erlanger, who shared a prize in Physiology or Medicine in 1944 (Chapter 6) came to play a special role in securing a prize for the only woman included among the 76 laureates recognized by the first 50 prizes in this discipline. This was Gerty T. Cori, who together with her husband Carl F. Cori received half a prize in 1947 "for their discovery of the course of the catalytic conversion of glycogen." She was nominated only twice, in 1946 and 1947, in both cases by Erlanger. In cases when recipients decide to share their prize money with co-workers (the insulin example in Chapter 6), this may be an indication of an unwarranted exclusion. George H. Whipple, who was one of three prize recipients in Physiology or Medicine in 1934 "for their discoveries concerning liver therapy in cases of anaemia," decided to share his prize money with his collaborators. In particular, he recognized his debt to his co-worker Frieda Robscheit-Robbins[6].

To conclude this section discussing radio astronomy it can be mentioned that "serendipity" has spawned a befitting acronym: in 1979 the University of California started a project entitled Search for Extraterrestrial Radio Emissions from Nearby Developed Intelligent Populations — condensed to SERENDIP.

The "vital force" of Nature dismissed

Chemistry, like any branch of science, includes many examples of serendipitous discoveries throughout its history: one of them actually led to the establishment of the whole field of organic chemistry. In the early 19th century it was believed that the chemistry of molecules of life is unique and depends upon a special "vital force" inherent in the molecules involved. In the 1820s, Friedrich Woehler, a dedicated chemist, went to Stockholm to train with Jöns Jacob Berzelius, the father of the organic chemistry concept. Back again in his laboratory in Berlin, Woehler performed an experiment that made him famous. He sought to prepare ammonium cyanate from two typical inorganic salts, potassium cyanate and ammonium sulfate, and to his surprise obtained crystalline urea, an organic substance. This result eventually led to the understanding that organic chemistry is nothing but a special branch of general chemistry. This branch is today referred to as the chemistry of carbon compounds.

Fermentation has been used for the processing and preservation of food throughout millennia of human civilization. Winemaking is known to have taken place as early as 7000–8000 years ago, starting in the Caucasus area and the Zagros mountains in present-day Georgia and Iran. Beer brewing and fermentation of milk followed some thousand years later. The first use of leavened bread stems from ancient Egypt, about 1500 BC. In chemical terms, the process of fermentation generally means a breakdown of sugars to alcohol and carbon dioxide by use of yeasts or bacteria, in the absence of oxygen. It was studied extensively by Louis Pasteur, who concluded that fermentation is catalyzed by a vital force, which he referred to as "ferments"; he implied that only intact cells could generate ferments.

It took a long time before the true nature of fermentation was clarified[7]. After numerous unsuccessful attempts to prepare active yeast extracts that would produce alcohol from sugar, Eduard Buchner got involved, in the 1890s. Eduard's older brother Hans, who was inspired by Robert Koch's work (see Chapter 6), had the idea that perhaps extracts of microorganisms could produce substances of use in medicine. In these studies he turned to yeast, but experienced a technical problem: this organism was tough to disintegrate. An assistant came up with the idea of grinding with quartz sand and diatomaceous earth, and of retrieving the extracted fluid by putting it under pressure. Another problem then surfaced: this extract did not store well. Hans tried to solve this problem by using the classical method of adding high concentrations of sugar, as a means of preservation. At this phase of experimentation,

Eduard Buchner (1860–1917), recipient of the Nobel Prize in Chemistry 1907. [From *Les Prix Nobel 1907.*]

his brother Eduard was visiting on a vacation from Tübingen. He observed that the addition of sugar led to a sparkling formation of gas in the mixture. From this he — but not his brother Hans — drew the momentous conclusion that the yeast extract by itself had the ability to ferment sugars, with the formation of alcohol and carbon dioxide. By use of several controls he could exclude the possibility that intact cells had remained in the preparations. This discovery made him famous overnight, and he became the first organic chemist to receive the Nobel Prize in Chemistry, in 1907. The motivation was "for his biochemical researches and his discovery of cell-free fermentation." His discovery eventually put an end to the belief in a "vital force" associated with intact cells. Ten years later he died from wounds received in action at the front in World War I.

Helpful students' mistakes

In the 1990s, the Japanese scientist Hideki Shirakawa was studying ways of synthesizing polyacetylene, with the aim of controlling the proportions of the cis- and trans-isomers of the polymer. The mixtures he obtained formed a disappointing black film on the inside of the reaction vessels. Once, one of his students by mistake added 1000 times more than the intended amount of one of the reagents, and to Shirakawa's surprise, a beautiful silvery film had developed. This incident put the researchers on a new track. The silvery

From left: Alan J. Heeger, Alan G. MacDiarmid (1927-2007), Hideki Shirakawa, recipients of the Nobel Prize in Chemistry 2000. [From *Les Prix Nobel 2000.*]

film contained only trans-polyethylene and by modifying the temperature, they soon succeeded in producing another, now copper-colored film composed only of cis-polyethylene. But there is more to this serendipitous story. At the same time that Shirakawa made his discovery, the chemist Alan G. MacDiarmid and the physicist Alan J. Heeger in the USA were experimenting with metallic-looking films containing the inorganic polymer sulphur nitride. MacDiarmid referred to his kind of film in a lecture in Tokyo, and during a coffee break he bumped into Shirakawa, who had come to the same meeting to talk about his metallic films. A collaboration between the three researchers ensued that led to the dramatic finding that iodine-doped trans-polyethylene showed a ten million times increased electrical conductivity! In 2000, Shirakawa together with Heeger and MacDiarmid received the Nobel Prize in Chemistry "for the discovery and development of conductive polymers." Such is the progress of science, its course erratically influenced by serendipitous events and accidental encounters.

From observational to evidence-based medicine

The infectious nature of some diseases in man and animals was recognized already in the early phase of settled human civilization. However, it took a long time before microorganisms and viruses were identified as causative agents (see Chapters 3 and 6). In spite of this, a highly successful prophylaxis against smallpox was introduced as early as the end of the 18th century. Edward

Jenner, a practising country doctor in Gloucestershire, was puzzled by the observation that milkmaids had nice smooth facial skin, whereas other girls' and ladies' faces were marred by pockmarks. He deduced that an infection with cowpox, frequently seen in milkmaids, might be the source of protection against smallpox. It would take about another hundred years before smallpox virus and the nature of immunity after infection would be characterized, but it was Jenner who was the first to actively transfer cowpox material from an infected animal to man. In May 1796, he inoculated eight-year-old James Phipps with material from cowpox vesicles and found the boy resistant to subsequent inoculation (!) with material from a human case of smallpox. Since the material used for immunization came from a cow, Lat. *vacca*, the procedure was referred to as "vaccination." The very same principle, the Jennerian approach, resulted in the worldwide eradication of smallpox, which the World Health Organization was proud to announce in 1978. This remarkable success in preventative health care, which indeed fulfills the requirement in Nobel's will to "be of benefit to mankind," was not awarded with a prize in Physiology or Medicine. In principle, the will allows prizes to be given to organizations, but this is ruled out at the Karolinska Institute (see previous chapter). In addition, the success of smallpox eradication was due more to organizational skills than to a recent single scientific discovery. The successful outcome of the eradication campaign is indeed a triumph for global international collaboration. Could this kind of contribution be recognized by a Peace prize?

The paradigm example of a serendipitous finding?

A story told many times over is the discovery of penicillin following an accidental observation made by Alexander Fleming (see also Chapter 6). In the early 1920s, Fleming was engaged in testing different antiseptics on bacteria and white blood cells. When he suffered from a severe upper respiratory infection, he managed to isolate some bacteria and to grow them on agar in a Petri dish. As he examined the dish, a teardrop fell on the agar, and when inspecting the plate one day later, he observed a clear space around the spot where the tear had fallen. Clearly, bacterial growth had been prevented, and further studies revealed that tears contain an anti-bacterial substance, which he later named lysozyme. However, this substance was not of practical use since it mainly prevented the growth of harmless bacteria.

In the summer of 1928, when Fleming returned from a vacation, he again noticed a bacterial plate with a clear circular area devoid of microbial growth.

In its center he noticed a mold, the spore of which had somehow entered the dish. Because of his previous experience with lysozyme, he did not discard the contaminated plate, but set out to characterize the mold and the substance it produced. The mold turned out to be of a rare kind possibly coming from a laboratory on the floor below, where a conceivable relationship between mold exposure and asthma was studied. He was not very successful in his endeavor, to identify the active substance and the project stalled. Fleming did, however, preserve the plate for the future by exposing it to formalin. It can be seen today in the British Museum in London.

After a considerable time, the project was picked up by other researchers and — as discussed in Chapter 6 — the active substance could be produced in a purified form and in quantities that could be used in patients. Once it became available, it came to save millions of lives. Alexander Fleming, Howard W. Florey and Ernst B. Chain shared the Nobel Prize in Physiology or Medicine in 1945.

Gerhard Domagk (1895–1964), recipient of the Nobel Prize in Physiology or Medicine 1939. [From *Les Prix Nobel 1939*.]

Before that, in 1939, the College of Teachers at the Karolinska Institute had awarded a prize to Gerhard Domagk for his finding that sulphonamides can kill bacteria. The prize was awarded "for the discovery of the antibacterial effects of prontosil." This discovery, in fact, came from a misconception. The simple hypothesis of the time was that bacteria are composed of proteins — which certainly is not the whole story — and might be attacked by certain dyes, specifically those containing a sulphonamide group. The German pharmaceutical industry had found that these would stick particularly firmly to protein, and dyes interfering with the growth of bacteria were identified. However, it turned out that their mechanism of action was not due to the part of the molecule that makes it a dye. Instead, it is associated with the sulphonamide group, the part attaching to the bacteria. As with the discovery of other effective antibacterial drugs, there are fascinating stories about their first successes in human patients. One is about Domagk in desperation giving the compound to his daughter when she was suffering from a serious streptococcal infection — the treatment leading to her miraculous recovery. When Domagk was informed that he had been selected for the prize in Physiology or Medicine, he wrote a letter of acceptance to the Karolinska Institute, but soon thereafter was forced by the Gestapo to decline. The Nazis also prohibited his visit to Stockholm, the reason being that Hitler was so annoyed when the Peace prize was awarded to Carl von Ossietzky in 1936 that he did not allow any German citizen to accept a Nobel Prize of any kind. Only after the war, in 1947, could Domagk come to Stockholm, to give his Nobel lecture and receive his medal and diploma. The prize money, however, had at that time been returned to the Nobel Foundation.

The discovery of infectious agents in a remarkable context

The first time I used the word "serendipity" in public was on December 10th, 1976, when I introduced the recipients of the Nobel Prize in Physiology or Medicine, Baruch Blumberg and Carleton Gajdusek. Both their discoveries — quite different from each other — were praised for implying "the introduction of new principles in the field of infectious diseases." Who would have thought that what at first appeared to represent a blood group antigen, the Australia antigen, would eventually be identified as the circulating surface antigen of hepatitis B virus in a chronic infection? And who would have thought that the ritual cannibalism practiced by the Fore people in New Guinea during and before the 1950s was the mechanism of transmission of a

Baruch Blumberg, recipient of one half of the Nobel Prize in Physiology or Medicine 1976. [From *Les Prix Nobel 1976.*]

highly atypical infectious agent, eventually called "prion," to individuals who at a much later date developed the lethal neurological disease kuru? These were indeed serendipitous discoveries! Chapter 8 gives a comprehensive account of the remarkable developments in what came to be called the prion field; this domain of research has been recognized not only by one half of the Nobel Prize to Gajdusek, but also to Stanley B. Prusiner alone in 1997.

The development of Blumberg's work that led to identification of the virus causing hepatitis B deserves a more thorough description. As will be described in Chapter 4, the large-scale use of the live yellow fever vaccine — a discovery awarded a Nobel Prize in Physiology or Medicine in 1951 — during World War II led to more than 300,000 cases of jaundice among the roughly 6.4 million soldiers who were immunized. It took a long time before the vaccine contaminant was identified — it turned out to be a virus that may cause chronic infections and contaminate blood and blood products. It was identified after another virus that also causes a liver infection, and hence named hepatitis B virus. The antecedent hepatitis A virus is a small poliovirus-like agent that infects the intestinal tract and spreads by fecal-oral transmission. The nature of hepatitis B virus remained enigmatic for a long time, a situation that changed only after Baruch Blumberg entered the field. His approach was fortuitous, and his solution to the problem came through a strange indirect approach[8].

Blumberg comes from a Jewish family. As emphasized in the previous chapter, scientists with this ethnic background are impressively over-represented as prize recipients in all three science disciplines. After an initial training in mathematics, he studied medicine at the College of Physicians and Surgeons of Columbia University, New York, where he became interested in global medicine and the health problems of peoples in developing countries. He started to travel widely to provide medical services and to collect blood samples. His first clinical training was at the Arthritis Division of Columbia University, and from then on, most engagements were in research on arthritis and in population genetics. He did his PhD at Oxford, UK, studying hyaluronic acid, a substance of importance in arthritis. His further professional development took place at the National Institutes of Health, Bethesda, and since 1964 at the Institute for Cancer Research, later called the Fox Chase Cancer Center in Philadelphia. Blumberg continued to collect blood samples from many populations worldwide, and he analyzed them in various cross-wise tests with the objective of detecting new antigens and antibodies in serum. Novel antigens representing physiological body products were detected in these assays, and their heritability was evaluated. One antigen was present in a particularly high frequency among Australian aborigines, and was therefore referred to as Australia antigen (Au antigen for short). It proved difficult to nail down the genetic determinants governing the presence or absence of this antigen. Various patient groups were studied, which resulted in the finding of the Au antigen at high frequencies in children with Down's syndrome, in individual recipients of blood transfusions, and also in a few particular cases of hepatitis. It eventually dawned on the scientists that the antigen they were studying might not be a physiological body protein, but perhaps a virus.

Blumberg was a newcomer to virology, with no training as a microbiologist. So he and his colleagues started to study textbooks on viruses, in order to understand this kind of agent. Their first attempt to publish data suggesting that Au antigen might be associated with hepatitis B was rejected. In extended studies, they substantiated their hypothesis and published data to support it in 1967; using electron microscopy, they eventually identified small globular particles as representing the antigen. These particles did not contain nucleic acid, which led to wild speculations that the protein alone was responsible for infectiousness. In this case, the hypothesis was incorrect, whereas in the case of prions (Chapter 8) it turned out to be true. It was not until 1979 that the virion structure was demonstrated by electron microscopy, whereas the Au antigen was found to represent isolated surface building blocks of the much larger virus particles.

The discoveries had many practical consequences. Now carriers of hepatitis B virus could be detected and excluded from serving as blood donors, thereby preventing the spread of the infection by transfusion. Furthermore, the particles that Blumberg and collaborators had discovered could be used to develop a vaccine. Antigen was isolated from the blood of persons carrying a hepatitis B infection in a way that excluded infectious viruses. Later on this method was replaced by the use of antigen produced in the laboratory by employment of recombinant DNA techniques. Hepatitis B vaccines are now used globally to prevent infections, which has two important consequences: they ensure that vertical spread of the infection—critical to the survival of the virus in nature—from mother to child at birth is interrupted; their other impact is that the form of liver cancer, which may develop in hepatitis B virus carriers, will eventually be kept in check.

"Serendipity" — From arcane to en vogue

When Merton and Barber finished their manuscript on the *Travels and Adventures of Serendipity* in 1958, they had identified 135 public usages of the word in the 83 years since Edward Solly first brought it to light. In the Afterword of the book published 45 years later, Merton emphasized the strikingly increased use and popularity of the word in scholarly and scientific journals. In the 1950s it was used 44 times, while since 1980 usage has increased to over 200 times. In the public press it was even more dramatic: the word had escaped from the esoteric domains of antiquarians, bibliophiles and a handful of scientists with particular interest in it and become used extensively. Walpole's original definition became diluted to let "serendipity" mean any event that can result in a pleasant surprise. The review of a full-text database covering more than 18,000 sources of mainly newspapers and magazines showed an increase from two-fold between 1960 and 1969 to more than 13,000-fold between 1990 and 1999. The word appeared in the titles of 57 books, which includes a collection of lectures by Umberto Eco, simply entitled *Serendipities*. In 2002, the romantic comedy *Serendipity* starring John Cusack and Kate Beckinsale hit the movie theatres. "Serendipity" was the title of a catalogue advertising women's underwear and a name of restaurants, inns and shops for arts and crafts. In 2000, it was the tenth most popular name for boats in the USA. The same year, the London Festival of Literature and Bloomsbury Press arranged a self-selecting poll asking for choices of "favorite words" in the English language. "Serendipity" headed the list, and the competition was probably fierce since "Jesus" and "money" tied for the tenth place. On October 18, 2001, "serendipity" was the word of the day. Anyone interested in getting an impression of its popularity may want to make a Google search on the World Wide Web. The response is overwhelming: in October 2009 5,770,000 hits were registered.

What's in a word?

When Horace Walpole coined his "expressive word," he had many options. Alternatives could have been "serendippery," "serendiption" or "serendipness" — but for some idiosyncratic reason, he settled for "serendipity," a fortunate choice in view of the term's appeal to philologists and poets. It has been referred to as "sweet and insinuating," its sound has been called "attractive," "euphonious." This is nicely illustrated in the poem "Serendip and Taprobane," by Anne Arwood Dodge published in 1927:

> *Words as argent-chimed as rain,*
> *Words like little golden beads,*
> *Apple and pomegranate seeds,*
> *Strung upon a silver thread,*
> *Little drops of lacquer-red,*
> *Tintinnabular and sweet,*
> *Little words with crystal feet*
> *Running lightly through my mind.*
> *If my lazy wit could find*
> *Gilded phrases to express*
> *Their perfected loveliness,*
> *I could make a cage of words*

Where, like bright heraldic birds,
They should strut and flaunt and preen,
Scarlet, silver, gold and green,
Elegantly strange and vain —
Serendip and Taprobane.

Taprobane, like Serendip, is an obsolete name for Sri Lanka. One wonders what would have happened if Horace Walpole had coined "taprobanity" instead of "serendipity." Would it have become as popular?

Linguists have suggested several factors that might contribute to the pleasing effect of "serendipity." One concerns the balance between the consonants towards the end of the word — with the stress on *dip* it is easy to pronounce. Another factor is the repetition of vowels, the two *e*'s followed by the three *i*-sounds. Finally, it is the symmetrical pattern, with two unstressed syllables surrounding the central emphasized syllable *dip*. Also, to an English-speaking person the separate meaning of "dip" adds to the flavor of the word. It is noteworthy that Walpole wrote: "I find everything I want … , wherever I *dip* for it," when he introduced it. Is the word appreciated as appealing and euphonious also by individuals with native tongues other than English? During the second half of the 20th century "serendipity" has diffused into all dictionaries, even abridged ones, it has been imported into many languages and can be found written in, e.g., Cyrillic letters. Importing new words into languages that use logograms may be difficult, but an example for how this is solved can be found in Kenkyasha's New Japanese English Dictionary, 1954. Here the word is cleverly transmuted into *horidashi* — dig out, pick up, find (a treasure, lucky find); to fling out — and *jozu* — skilled, adept, expert.

The capricious nature of the scientific process

In this chapter it has been noted repeatedly that the creative scientific process is frequently ignited by luck striking the prepared mind. This has been well formulated not only by Louis Pasteur, but even earlier in the 19th century by Joseph Henry, the famous American scientist who became the first Secretary of the Smithsonian Institution. He said "The seeds of great discoveries are constantly floating around us, but they only take root in minds well prepared to receive them." It helps to be in the right place at the right moment and to possess a well-developed curiosity. But, the reader may ask, does science not follow a strictly logical and rational intellectual pattern? There is of course a form of science that implies the application of established techniques to well-defined problems. This science, which I have sometimes referred to as "horizontal science," has its own value, but it does not allow the quantum leaps, Thomas Kuhn's epistemological paradigm shifts[9], that cause dramatic changes in our perspective on knowledge. The major discoveries, made in the process of this "vertical science," require other qualities, only some of which we can define. Thus high quality science is not a predictable venture, and this is what gives the endeavors such a charm. On a few occasions I have been asked whether one can plan for serendipitous events. The answer is that one cannot. So the question remains how mentors should encourage students to develop their individual — their subjective creativity. There is no general recipe, I am afraid. What we need to encourage in them is to acquire a solid basic knowledge, without enforcing the belief that everything in textbooks is true. Also, to stimulate an openness of mind, allowing a build-up of confidence, permitting the formulation of even odd and lateral, possibly intuitive, hypotheses. This is sometimes referred to as thinking "out of the box." Curiosity needs to be given free rein. And of course we need to teach the capacity to test the various hypotheses experimentally, and to debate the outcome of the experimentation in the dialectic discourse that is an integral part of the scientific process. Finally, we may also infuse into our students some understanding for the unique nature of the scientific process, including references to its history and the role of serendipitous events.

Since the advance of science is unpredictable, how can we successfully apply for grant money to pursue our research? The answer is that the degree of unpredictability in scientific efforts varies, and with them the likelihood that a certain approach will give an answer to the problems posed. If in the pursuit of a project advances are made in an unforeseen direction, it is something the granting authorities need to accept and even enjoy. In their evaluation the emphasis therefore has to be on the conceptual and experimental quality of the work, and also on the track record of the applicant. An approach used by some applicants is to include proposals for experiments they have already performed; the results then are predictable.

An evaluation by scientific colleagues, the "peer review" system, is the generally accepted approach to evaluation of research grant applications. Provided the referees have the requisite competence, this system offers good opportunities to gauge the quality and novelty of different approaches. However, the referees may prefer to play safe and recommend support for projects with a predictable outcome. By definition, the chances for success of bold approaches derived from heterodox thinking cannot be anticipated, and hence there is a risk that they will not receive any support. One way of giving even the odd project a chance would be to put aside a small fraction of the available grant money for unconventional, maybe even "crazy" applications.

A final remark about how scientists by convention present their data in scientific publications, a problem discussed at length in the book by Merton and Barber[4], should be given. A non-scientist not only believes that the scientific process is rational; he also believes that a typical scientific article gives a proper presentation of the problem and an accurate picture of the sequence of events underlying a particular discovery. Nothing can be farther from the truth, and a number of Nobel Laureates have commented on this. George W. Beadle was one of the dominant scientists in the field of genetics (Chapter 7). Through his work in the 1940s and 1950s, he came to mark the transition from classical to molecular genetics. In 1958, he received one half of the Nobel Prize in Physiology or Medicine together with Edward L. Tatum. The other half of the prize went to Joshua Lederberg as described in Chapter 6. Beadle started a review paper[10] in 1966 with the statement: "I have often thought how much more interesting science would be if those who created it told how it really happened, rather than reporting it logically and impersonally, as they so often do in scientific papers."

In the same spirit Peter Medawar, the polymath biologist, who shared a Nobel Prize in Physiology or Medicine with Sir Frank Macfarlane Burnet in 1960 for revolutionizing contributions to the field of immunology, even entitled a lecture on BBC television "Is the Scientific Paper a Fraud?" Of course he did not mean to imply that the facts were incorrectly presented but rather that the process of thought and all the failed experiments could not be taken from the presentation. The inimitable physicist Richard Feynman in his 1965 Nobel Prize address elaborated on the same theme[11]. His lecture starts: "We have a habit in writing articles published in scientific journals to make the work as finished as possible, to cover up all the tracks, to not worry about the blind alleys or to describe how you had the wrong idea first, and so on. So there isn't any place to publish, in a dignified manner, what you actually did in order to get to do the work … ." And the first paragraph finishes: "So, what I would really like to tell you about today are the sequence of events, really the sequence of ideas, which occurred, and by which I finally came out the other end with an unsolved problem for which I ultimately received a prize."

Another quote in the same spirit is from Roald Hoffmann, the laureate in Chemistry in 1981 and a qualified poet. In an article published in a regular scientific journal[12] he wrote: "In order to present a sanitized paradigmatic account of a chemical study, one suppresses many of the truly creative acts. Among these are the "fortuitous circumstance" — all of the elements of serendipity, of creative intuition at work."

This chapter began by mentioning that not only Walpole but also Voltaire had been impressed by the oriental tale of the princes from Serendip, which he incorporated in episodes of his novel *Zadig*. In that novel he reflects upon how, throughout the flow of our daily lives, we are exposed to happy — and unhappy — accidents and encounters. This theme was also reviewed in the book by Merton and Barber[4], but Merton further developed it as "the unanticipated consequences of the purposive social action" in some of his other influential books in social sciences.

If we can develop the capacity to make the proper intuitive and sagacious choices, we can enrich our individual lives. Thus there are parallels between the pursuits in our private lives and the engagement of scientists generating new, reproducible knowledge. But there are also major differences. In science, the goal is to search for the "truth," the correct and comprehensive understanding of Nature. These endeavors have led to a progressive in-depth understanding of very complex and coherent systems; the accumulated knowledge has allowed us to develop powers by which we markedly influenced the conditions of our lives. Our modern society is based on the advances made by science and technology. Scientific achievements may also allow us to make some predictions.

In our personal lives, we make a number of choices that may set us on irreversible tracks. This has been well described in the poem *The Road Not Taken* by Robert Frost — "and that has made all the difference." Fortuitous and unpredictable events play a major role in our private lives, and may also have a decisive importance in our performance as scientists, or as artists for that matter. Their outcomes are unpredictable. If one has the privilege to be a scientist, the two unpredictable sequences of events may intertwine in a way that allows for a rich, exciting — and serendipitous — life, which I can testify to from my own experience.

Coda — Minute infectious agents

The discovery of bacteria by Louis Pasteur and Robert Koch (Chapter 6) prompted the development of means to sterilize fluids. One approach was to push the liquid infectious material through porcelain filters which had pores so small that bacteria could not pass through. For unknown reasons, perhaps simply curiosity, a Russian scientist Dmitri I. Ivanovsky took fluid from tobacco plants infected by an agent causing a "mosaic" disease in them and passed it through porcelain filters. To his surprise he found that the infectious property remained in the filtrate. He was himself not convinced of his finding and speculated that there might be cracks in the filters or that the agent might appear in a minute spore form. The definite proof that there were infectious agents smaller than bacteria was provided by the Dutch scientist Martinus W. Beijerinck. He repeated Ivanovsky's experiment and concluded that he had found a new type of small replicating agents, which he called *virus*. Since he could not comprehend that such a small thing as a virus could be particulate he called it *contagium vivum fluidum*, the "floating contagious form of life." It took some 60 years before the true nature of viruses was understood and the development of this knowledge will be the theme of the next chapter.

Chapter 3

Nobel Prizes and the Emerging Virus Concept

Sven Gard (1905–1998), professor of virology at the Karolinska Institute (1948–1972). [Courtesy of the Karolinska Institute.]

In the fall of 1958 I was an early third year medical student and came to listen to lectures on viruses by Professor Sven Gard at the Karolinska Institute in Stockholm. His engaging presentations changed my life. I immediately started to do virus research in parallel with my studies and became a researcher and teacher instead of a clinically active physician as originally planned. However, little did I know that the presentation of the nature of viruses Gard gave derived from very recently consolidated concepts about the unique features of this category of infectious agents.

One picture that Gard drew on the blackboard showed that the size distribution of bacteria, representing the variations expected to be found in cells that grow and divide, contrasted with the homogenous size distribution of virus particles. He also described the two-phase life cycle of viruses. Their extra-cellular transport form, with the genetic material enwrapped in a protein shell and in some cases additional structures, was contrasted with the reoriented metabolism of the infected cell leading to the production of different components eventually assembled into new virus particles. In my own early career I could follow how the virus concept consolidated further and from 1960 and onward I had opportunities to learn to know the Nobel Laureates in this field of medicine.

The development of the modern concept of a virus evolved slowly over more than 50 years. There are a number of articles and books that have described the critical steps in this development[1-7]. It was through the pioneering work of Robert Koch and Louis Pasteur that bacteria, visible by the light microscope, were found to be the cause of many infectious diseases. However, as early as the 1880s, while working with rabies, a disease later on demonstrated to be caused by a virus, Pasteur reflected on the agent as being a "microorganism infinitesimally small." A few years later Koch expressed concerns about the fact that the cause of many infectious diseases remained undefined. He gave as specific examples influenza and yellow fever and speculated that they might belong to a quite different group of microorganisms.

Studies in a completely different context demonstrated that a disease in tobacco plants causing mosaic patterns to appear in their leaves was infectious. In 1882 it was demonstrated by Adolf Mayer in the course of experimental plant work at Wageningen, The Netherlands, that water extracts of leaves with the mosaic changes could transfer such changes to healthy plants. This observation prompted further studies of the tobacco mosaic agent, as it came to be called, and prompted the experiments, mentioned in the coda of the previous chapter, defining a new category of *ultrafiltrable* infectious agents. It was Beijerinck, who introduced the name *virus*. This name had been used through antiquity as a general term for phenomena that are unpleasant and dangerous. The meaning is wide-ranging, encompassing disease agents, vicious fluids from plants and secretions that have a bad odor or stench and poisons from animals. The Latin word virus, often referred to as toxin, probably has a Sanskrit origin and is related to the English word weasel, a member of the family of mustellides. Many members of this family have a capacity to spray strong smelling fluid from their peri-anal glands when threatened.

The first virus causing disease in animals was identified in 1898 by Friedrich Löffler and Paul Frosch in Germany. They demonstrated that the agent causing foot-and-mouth disease in cattle could pass through filters retaining bacteria. William H. Welch at Johns Hopkins University, Baltimore, Maryland, brought this observation to the attention of Walter Reed and James Carroll in the US who, four years later, demonstrated the first virus infecting humans, the yellow fever agent (Chapter 4).

The final development of the virus concept is closely intertwined with the development of molecular biology, which has been well presented in many books[8,9] and is further discussed in Chapter 7. André Lwoff's definition of viruses in 1957[10] is often referred to as the first comprehensive and resilient description of their nature. His description was:

> ... infectious, potentially pathogenic, nucleo-proteinic entities possessing only one type of nucleic acid, which are reproduced from their genetic material, are unable to grow and to undergo binary fission, and are devoid of a Lipmann system.

At this time it was eventually understood that "their genetic material" referred to the nucleic acid they contained. The last part of the statement notes that viruses have no energy-converting systems and generally very limited metabolic functions. However, they do have the qualities of life associated with replication and capacity to show genetic variation.

André Lwoff (1902–1994), shared the Nobel Prize in Physiology or Medicine 1965 with François Jacob and Jacques Monod. [From *Les Prix Nobel 1965*.]

Nobel Prizes in the field of virology

There have been a number of Nobel Prizes directly awarded for discoveries in the field of virology proper as well as in related fields. It started in 1946 when one half of the prize for Chemistry was shared by John H. Northrop and Wendell M. Stanley (Table 1). Stanley had managed to produce crystalline structures containing tobacco mosaic virus (TMV)[11], the virus that came to

Table 1. Nobel Prizes in Chemistry for research on Tobacco Mosaic Virus.

Year	Awardee(s)	Motivation
1946	John H. Northrop Wendell M. Stanley	"for their preparation of enzymes and virus proteins in a pure form" (½ prize)
1982	Aaron Klug	"for his development of crystallographic electron microscopy and his structural elucidation of biologically important nucleic acid-protein complexes"

play a particular role throughout the history of virology[12-14]. Much later another prize for Chemistry was given to Aaron Klug. A considerable part of Klug's work concerned the fine structure of TMV and this project was an extension of the prematurely disrupted pioneering crystallographic work by Rosalind Franklin at Imperial College, London, during the mid 1950s[15]. The remaining prizes in the field have not surprisingly been given in Physiology or Medicine. These prizes can be divided into different categories. Six prizes have been given for discoveries concerning animal viruses and the special kind of ultrafiltrable agents called prions, discussed in Chapter 8 (Table 2). Still another four prizes have been given for genetic studies using viruses infecting bacteria, bacteriophages or phages for short (Table 3). The term phage comes from Greek *phagus* "to eat." Finally there is a group of five prizes given for discoveries of general molecular biological principles involving studies in which various animal and plant viruses were used (Table 4). Examples of such discoveries are reverse transcriptase, cellular oncogenes and split genes.

Ever since 1948 and into this century the discipline of virology has been personally represented among the five members of the Nobel Committee (normally limited to a six-year appointment) or among adjunct members.

Table 2. Nobel Prizes in Physiology or Medicine for studies of animal viruses and prions.

Year	Awardee(s)	Motivation
1951	Max Theiler	"for his discoveries concerning yellow fever and how to combat it"
1954	John F. Enders Thomas H. Weller Frederick C. Robbins	"for their discovery of the ability of poliomyelitis viruses to grow in cultures of various types of tissue"
1966	Peyton Rous	"for his discovery of tumour-inducing viruses" (½ prize)
1976	Baruch S. Blumberg D. Carleton Gajdusek	"for their discoveries concerning new mechanisms for the origin and dissemination of infectious diseases"
1997	Stanley B. Prusiner	"for his discovery of prions — a new biological principle of infection"
2008	Harald zur Hausen Françoise Barré-Sinoussi Luc Montagnier	"for his discovery of human papilloma viruses causing cervical cancer" (½ prize) "for their discovery of human immunodeficiency virus" (½ prize)

Table 3. Nobel Prizes in Physiology or Medicine for genetic studies using bacteriophages.

Year	Awardee(s)	Motivation
1958	Joshua Lederberg	"for his discoveries concerning genetic recombination and the organization of the genetic material of bacteria" (½ prize)
1965	François Jacob André Lwoff Jacques Monod	"for their discoveries concerning genetic control of enzyme and virus synthesis"
1969	Max Delbrück Alfred D. Hershey Salvador D. Luria	"for their discoveries concerning the replication mechanism and the genetic structure of viruses"
1978	Werner Arber Daniel Nathans Hamilton O. Smith	"for the discovery of restriction enzymes and their applications to problems of molecular genetics"

Table 4. Nobel Prizes in Physiology or Medicine in which animal or plant viruses were used in discoveries of basic biological principles.

Year	Awardees	Motivation
1975	David Baltimore Renato Dulbecco Howard Temin	"for their discoveries concerning the interaction between tumour viruses and the genetic material of the cell"
1989	J. Michael Bishop Harold E. Varmus	"for their discovery of the cellular origin of retroviral oncogenes"
1993	Richard J. Roberts Phillip A. Sharp	"for their discoveries of split genes"
1996	Peter C. Doherty Rolf M. Zinkernagel	"for their discoveries concerning the specificity of the cell mediated immune defence"
2006	Andrew Z. Fire Craig C. Mello	"for their discovery of RNA interference — gene silencing by double-stranded RNA"

Since 1970 the committee has been composed of five ordinary members and ten adjunct members. This enlargement has been deemed necessary to cover the wide and diversified field of medical sciences. Until 1972 Gard carried the responsibility of representing virology and hereafter it was the privilege to serve in this function. When I left to become permanent secretary of the Royal Swedish Academy of Sciences in 1997, the field had already for some years been represented by one other virologist. It was the highly respected Finnish colleague

Ralf F. Pettersson, professor of molecular biology and Director of the Ludwig Institute for Cancer Research at the Karolinska Institute. He was an adjunct member from 1990–1994 and 2001–2005 and in the intervening period he was a full member of the committee, chairing it from 1998–2000.

Gard was the professor of virus research at the Karolinska Institute between 1948 and 1972. The chair in virology that he occupied is probably one of the oldest ever established. He was very influential in the work of the Nobel Committee at the Institute. Gard's first major engagement concerned evaluations that led to the award of a Nobel Prize to Max Theiler in 1951[16]. As will be discussed in the next chapter he carried out the critical examinations of Theiler's work only to be sidelined in 1951 by the chairman of the committee and Vice-Chancellor of the Institute, Hilding Bergstrand, a professor of pathology. During 1952–1954 Gard made seminal evaluations that led to the award of a prize in 1954 to John F. Enders, Thomas H. Weller and Frederick C. Robbins for their success in growing poliomyelitis viruses in cultures of various types of tissue[17] (Chapter 5). On this occasion Gard gave the presentation speech at the prize ceremony.

F. Macfarlane Burnet (1899–1985), recipient of one half of the Nobel Prize in Physiology or Medicine 1960. [From *Les Prix Nobel 1960*.]

The next time that Gard introduced Nobel Laureates was in 1960. In that year Frank Macfarlane Burnet and Peter Medawar received a prize "for discovery of acquired immunological tolerance." Burnet, besides his selective and impressive contributions to the field of immunology, was recognized pre-eminently as one of the giants in the field of virology to which he had made a number of different fundamental contributions. He had, therefore, since 1948 been nominated every year for discoveries made in his virus research. In 1950 it was in fact proposed by 4 dissenting members out of the 13-man committee, that Burnet should share a prize with Theiler. The proposed motivation for Burnet's contribution was "for his discoveries of methods that make cells resistant to certain virus infections." However the College of Teachers at the Institute followed the recommendation of the majority of the committee and gave a prize for the recently discovered hormones of the adrenal gland as discussed in Chapter 6. On the occasion of Burnet's visit to Stockholm to receive his prize I was invited to Gard's house for a formal dinner on December 8. It was an unforgettable experience for me as a 23-year-old "budding virologist," to use the host's formulation, to meet one of the giants in the field I had just entered.

Throughout the 1960s Gard introduced two more groups of prize recipients, in 1965 and 1969. Both of them recognized advances in phage genetics. However, as early as 1958 a prize had been given in the field of microbial genetics, which included studies of bacteriophages for carrying over genetic material — *transduction* — from one cell to the other. The half prize given to Lederberg concerned the latter phenomenon (Chapter 6). He was introduced together with his co-laureates George W. Beadle and Edward L. Tatum at the prize ceremony by Torbjörn Caspersson, a professor of cell biology and genetics at the Institute (Chapter 7). In 1965 the prize in Physiology or Medicine was given to François Jacob, André Lwoff and Jacques Monod. Lwoff's description of the phenomenon of lysogeny — the presence of dormant virus genetic material in a cell — was important in the understanding of the nature of viruses as we shall see.

In 1966, half the prize unexpectedly recognized the virologist Peyton Rous. This prize is remarkable in many ways. Rous discovered in 1911 that a virus was the cause of a certain kind of malignant sarcoma in chickens. This was not the first tumor virus that was identified. Already three years earlier the Danes C. Ellerman and O. Bang had described a virus causing leukemia in chickens. Rous was first nominated for the prize in 1926 and after that on repeated occasions. He was subjected to a number of evaluations. For example as early as 1939 another professor of pathology at the Institute, Folke Henschen, wrote: "Rous, who was born in 1879, has, since 1929, on repeated occasions been nominated for a Nobel Prize. The fact that for the last 14 years his name has belonged among those who are repeatedly nominated is remarkable in itself. Of even greater importance is perhaps that among the proposers there are many famous scientists like Landsteiner, Murray, Carrel and Meyerhof." In spite of Henschen's recommendation that he be seriously considered for a prize in that year and many similar recommendations later, Rous did not receive his prize until almost thirty years later at the age of 87, the record age for a laureate (page 31). He was introduced at the prize ceremony, not by Gard but by Georg Klein, a professor of tumor biology since 1957.

The founding fathers of the field of phage genetics, Max Delbrück, Alfred D. Hershey and Salvador D. Luria, eventually got a prize in 1969 and the prize citation refers to their studies of the replication mechanism and the genetic structure of viruses. After this year no more prizes were given for discoveries in phage genetics, but indirectly Werner Arber's contribution to the discovery of restriction enzymes was dependent on the use of phage systems. This discovery was rewarded in 1978 by a prize that he shared with Daniel Nathans and Hamilton O. Smith (Table 3). Other prizes after 1969 have focused on discoveries in studies of animal viruses and the connected subject of prions as ultrafiltrable agents of a non-viral nature (Table 2). The prize in 1976 to Blumberg and Gajdusek highlighted the unique pathogenesis of hepatitis B, which was discussed in Chapter 2, and the first identification of a human prion disease. The latter term was not introduced until later by Prusiner, who received his own prize in 1997[18]. The final chapter presents the amazing developments in the prion field.

Recently advances in studies of conventional viruses were again rewarded by a Nobel Prize in Physiology or Medicine. The 2008 prize went to Harald zur Hausen for his pioneering studies of papilloma viruses and their role in cervical cancer and to Françoise Barré-Sinoussi and Luc Montagnier for the discovery of human immunodeficiency virus (HIV). In the selection of only the two latter scientists as discoverers of HIV, the committee used a very narrow definition of discovery. The isolation of a retrovirus from patients with *Acquired ImmunoDeficiency Syndrome* (AIDS) was recognized, but not the definite identification of the nature of the virus and the development of the important first antibody test.

There have been five prizes in Medicine for discoveries of general molecular biological importance in which virus systems were used (Table 4). The prize in 1975 was given to David Baltimore, Renato Dulbecco and Howard M. Temin for studies of the interaction between tumor viruses and the genetic material of the cell, which included Baltimore's and Temin's discovery of reverse transcriptase (Chapter 7). The tumor promoting oncogenes were recognized in 1989 by the award of the prize to J. Michael Bishop and Harold E. Varmus, while splicing (trimming of information-carrying RNA — Chapter 7) was recognized by the award to Richard J. Roberts and Phillip A. Sharp. In 1996, Peter C. Doherty and Rolf M. Zinkernagel were recognized for their observation in studies using lymphocytic choriomeningitis virus that the cell-mediated immune response is host cell restricted. Finally the discovery of interfering ribonucleic acid (RNAi) to which attention was given by the 2006 prize to Andrew Z. Fire and Craig C. Mello also goes back to earlier observations in the field of virology. Clearly viruses are useful tools to unravel fundamental phenomena in Nature, not least in the way in which genetic information is stored and managed (Chapter 7).

Nobel archives and the evolving virus concept

The purpose of this chapter is to take a complementary perspective on the historical development of the virus concept by analyzing the archives of the Nobel Prizes in Chemistry and Physiology or Medicine. Since these archives remain secret for 50 years[19], at the time of writing in early 2010, materials from 1959 and previous years could be reviewed for scholarly purposes. This means that all the underlying documents for the half prize in Chemistry to Northrop and Stanley in 1946 by the Royal Swedish Academy of Sciences and material concerning these candidates for a prize in Physiology or Medicine at the Karolinska Institute could be fully reviewed. The archive materials for the prizes in 1951 and 1954 were reviewed separately and will be discussed in the following two chapters. None of them include discussions with direct bearings on the interpretation of the evolving virus concept, but there are other more recent archive materials of potential importance for this. In 1955 Berndt Malmgren, a professor of bacteriology at the Institute, made an evaluation of Seymour S. Cohen, Delbrück, Hershey and Luria and in 1957 Gard evaluated Lwoff. A further evaluation of Hershey together with the first time nominees Gerhardt Schramm and Heinz L. Fraenkel-Conrat was made by Klein in 1959. Finally, the archive material pertaining to the prize in Physiology or Medicine to Lederberg in 1958 (Chapter 6) including an extensive review by Klein contains some information of importance to the discussion below.

The first Nobel Prize in virology: Wendell Stanley

Stanley was nominated for a Nobel Prize in Chemistry for the first time in 1938 and subsequently for a further 13 times until he was awarded it in 1946 (Table 5). The basis for the nominations was his crystallization of TMV[11]. The original nominee Harold C. Urey referred to this spectacular finding and argued that Stanley had proven that the crystallized virus protein contains a "substance that is both a protein in the generally accepted sense and at the same time is the active virus causing the disease." He also stated toward the end of his nomination letter "that there is apparently no sharp dividing line between ordinary proteins, which are considered not to be living, and bacteria which are considered to be living organisms. In this way his work is

Wendell M. Stanley (1904–1971), shared one half of a Nobel Prize in Chemistry 1946 with John Northrop. [From *Les Prix Nobel 1946*.]

The (Theodor) Svedberg (1884–1971), recipient of the Nobel Prize in Chemistry 1926, and Arne W. K. Tiselius, recipient of the Nobel Prize in Chemistry 1948. [From *Les Prix Nobel 1926, 1948.*]

a notable contribution to the understanding of the relationships between dead and living substances." The proposal was subjected to a thorough evaluation by The Svedberg, a professor of physical chemistry, Uppsala University and recipient of a Nobel Prize in Chemistry in 1926. Stanley was then evaluated two more times by Svedberg and then finally by Arne Tiselius, a professor of biochemistry at the same university.

The choice of Svedberg and Tiselius as evaluators requires some discussion. By present-day standards they would be considered to be too close to Stanley to serve in this function. Clear conflicts of interest can be identified. Svedberg's work in Uppsala was heavily dependent on support from the Rockefeller Foundation, as was Stanley's work since he was employed at the Rockefeller Institute for Medical Research at Princeton. Stanley had sent material to Svedberg's laboratory (according to one source Svedberg picked it up himself) for determination of its molecular weight by ultracentrifugation[20]. Tiselius, who was one of Svedberg's foremost students, spent a year at the Rockefeller laboratory at Princeton in 1935 on a scholarship from the Foundation and was also dependent on support from it. Tiselius himself received a Nobel Prize in Chemistry in 1948. On a personal note I would like to stress that impartiality, objectivity and secrecy always have been unexpressed founding stones of the code of ethics guiding work in Nobel Committees. During my twenty years of association in different functions with the committee at the Karolinska Institute we did not have to declare our affiliations with the individuals we were evaluating, but at a later time a requirement to declare background information pertinent to potential conflicts of interest was introduced.

Table 5. Nominations of Wendell Stanley for a Nobel Prize in Chemistry and the evaluations.

Year	Nomination(s)	Evaluator	Recommendation
1938	1	The Svedberg, professor of physical chemistry, Uppsala University	Delay decision
1939	2*	The Svedberg	Delay decision
1941–44	6	No evaluations	
1945	2	The Svedberg	Prize-worthy
1946	2	Arne Tiselius, professor of biochemistry, Uppsala University	Delay decision

*One nomination by Svedberg also included Frederick C. Bawden and Norman W. Pirie.

The conclusions of the evaluations of Stanley were generally equivocal and a delayed decision was proposed (Table 5). A possible combination with the British plant virologists Frederick C. Bawden and Norman W. Pirie (Svedberg himself nominated these three scientists in 1939) was discussed. It was also considered if a prize in Physiology or Medicine would be more appropriate. Only in 1945 did Svedberg give a direct recommendation for a prize in Chemistry. In the following year Tiselius made a final review of Stanley, now together with James B. Sumner and John H. Northrop, the head of the institute where Stanley was working. Tiselius came to the conclusion that Sumner and Northrop should be considered for the prize the same year but that a prize for Stanley should wait, partly due to the fact that other candidates in virus biochemistry needed to be considered. This hesitation was recaptured in the written deliberations by the committee. Due to limitations of the state of chemical virus research at the time it was stated that there are "uncertainties about the conditions of the virus in the living cell." However, in the final discussions in the committee Tiselius apparently changed his mind and the committee decided, somewhat surprisingly, to divide the prize and give half of it to Sumner "for his discovery that enzymes can be crystallized" and the other half jointly to Northrop and Stanley (Table 1). It is difficult to understand why the prize was not divided differently with one half to Sumner and Northrop and the other half to Stanley. Or would this have given too much emphasis to Stanley's work? One may further ask why Stanley was chosen for this prize in Chemistry and not one in Physiology or Medicine. What were the views of the committee at the Karolinska Institute when these decisions were made?

Table 6. Nominations of Wendell Stanley for a Nobel Prize in Physiology or Medicine and the evaluations.

Year	Nomination(s)	Evaluator	Recommendation
1937	1	Hilding Bergstrand, professor of pathology, KI	Prize-worthy
		Einar Hammarsten, professor of chemistry and pharmacy, KI	Not prize-worthy
1938	2	The Svedberg, professor of physical chemistry, UU	Prize-worthy
1939	7	Einar Hammarsten	Not prize-worthy
		Folke Henschen, professor of pathology, KI	Delay decision
		The Svedberg	Prize-worthy
1940–44	14	No evaluations	
1946	8	Folke Henschen	Delay decision
		Torbjörn Caspersson, professor of medical cell research and genetics, KI	Prize-worthy (?)

Abbreviations: KI — Karolinska Institute, UU — Uppsala University.

A comparison of the nominations of Northrop, Stanley and Sumner for a prize in Chemistry or in Physiology or Medicine with regard to numbers, timing and proposer is quite revealing (Tables 5 and 6). In assessing these data it should be kept in mind that the years 1937–46 include WWII; no prizes were given during three years and for two other years the prize was reserved and given with a delay of one year. Stanley was nominated earlier and much more frequently for a prize in Physiology or Medicine than in Chemistry, 32 times versus 13 (3 nominations from Swedes) times. Out of the first set of nominations six included Northrop and two of these also included Sumner.

Stanley was proposed for the first time for a prize in Physiology or Medicine in 1937. He was evaluated by Bergstrand and a professor of chemistry, Einar Hammarsten. It should be mentioned that Hammarsten came to have a long career as professor at the Institute (1928–1957) and was highly influential in the Nobel work and in the furthering of Swedish biomedical research[21]. We will meet him again in Chapters 6 and 7. Hammarsten was a pioneer in studies of nucleic acids and in 1938, together with Caspersson he showed the existence of DNA with a molecular weight exceeding one million. In his evaluation of Stanley, Bergstrand gave a thorough review of the history of virology and ended up enthusiastically endorsing the award of a prize to him. In contrast Hammarsten was not satisfied with the proposed chemical evidence for the protein nature of the infectious principle and declared Stanley's work as not worthy of a prize. This conclusion was repeated in another evaluation he made in 1939, which, however, praised the findings by Bawden and Pirie. In particular he emphasized that the latter scientists have shown that the virus contains nucleic acid, something that Stanley, because of his crude original chemistry, had overlooked. In total there are another 5 evaluations of Stanley's work (Table 6). Two of them that came from Svedberg endorse a prize in Physiology or Medicine, but the others are more hesitant about recommending a prize. Why was it so difficult for the medical committee to come to an agreement about what to do with Stanley? He was frequently proposed and in many cases by authoritative proposers.

Stanley's findings and the discussions of infectious proteins

As already mentioned TMV was the first infectious agent shown to pass through filters retaining conventional bacteria. Because this virus could be transmitted in a controled way between plants and be produced in large quantities and quantified by special assays on leaves, it became a favorite choice for virus studies. The complete molecular structural description of the virus particle as illustrated in Klug's Nobel lecture[15] was one concluding milestone in these developments. The whole story of TMV research is a very central part of the history of virology and has been presented in a number of publications[12-14] and Stanley's contributions have been summarized in a less flattering way[22]. He is described as a person adopting the intellectual program of others and to have been a relatively mediocre biochemist. His approach to crystallizing the virus was flawed by technical errors and the symmetrical aggregates contained impurities and water. They were not true three-dimensional crystals. X-ray diffraction studies in England by J. D. Bernal and I. Fankuchen showed that they were paracrystalline and that the rod-like virus particles, eventually defined, were randomly distributed in the direction of their long axis. However, Stanley's discovery did indeed stir the interest of the public press. He enthusiastically discussed how "dead" proteins could infect leaves and the newspaper headlines read "Life in the making" or similar formulations. Still, what is more problematic about Stanley's 1935 publication[11] is not that the aggregates are not true crystals, but that he completely missed the nearly 6% RNA in the virus particle, which was correctly described in 1936 by Bawden and Pirie[23]. He was quick to acknowledge his mistake, but wrote somewhat evasively in his Nobel lecture[24] that the nucleic acid "was reported by Pirie and co-workers in December, 1936, and *by the writer a few days later.*"

As already mentioned, Hammarsten in his evaluations concluded that the contribution in chemistry of Bawden and Pirie was prize-worthy, but not that of Stanley. What Stanley did show in later studies was the occurrence of some differences in amino acid composition of different strains of TMV, which he interpreted to represent mutations. TMV was the first virus to be visualized in the electron microscope and Stanley and his collaborators were quick to confirm these findings. It was Bernal and his group in the UK who first showed that individual virus particles had a symmetry indicating that they were composed of multiple units of proteins. The question then remains: what did Stanley teach us about the structure of viruses? In fact, it was not very much. His Nobel lecture[24] does not give much guidance.

He stated that when his work "was started in 1932, the true nature of viruses was a complete mystery. It was not known whether they were inorganic, carbohydrate, hydrocarbon, lipid, protein or *organismal* in nature." In another part of the speech he stated, "When viruses were found to possess the ability to reproduce and to mutate, there was a definite tendency to regard them as very small living organisms, despite the fact that the question of metabolic activity remained unanswered." He described the diversity of size and form of viruses known at the time and included in his presentation various animal viruses, plant viruses and also phages, "now regarded as viruses." The synthesis of the lecture is the statement, "The fact that with respect to size, the viruses overlapped with the organisms of the biologist at one extreme and with the molecules at the other extreme served to heighten the mystery regarding the nature of viruses. Then too, it became obvious that a sharp line dividing living from non-living things could not be drawn and this fact served to add fuel for discussion of the age-old question of *What is life?*"

Nowhere does he reflect on what happened inside the infected cell. He might have done that since Northrop, with whom he shared half of the prize, was heavily involved in this question[25]. Northrop himself surprisingly used *The preparation of pure enzymes and virus proteins* as a title for his Nobel lecture[26], perhaps to conform to the prize rationale given by the Academy. All he said about virus work in the lecture, besides mentioning that Stanley had crystallized TMV and Bawden and Pirie had done the same thing with bushy stunt virus of tomatoes, is that "A nucleoprotein which appears to be bacteriophage has been isolated but not crystallized." Throughout the 1940s

Northrop had a major disagreement with Delbrück about the formation of bacteriophages[25]. The importance of studies of this kind of infectious agent for the evolution of the virus concept will be discussed later. But before that it may be appropriate to let TMV return to the scene in a way that annuls the concept of an infectious protein.

In 1956 there was a nomination for a prize in Chemistry by Abraham Taub, Columbia University, New York, for a prize in Chemistry to Fraenkel-Conrat and Robley Williams, both at the Virus Laboratory of the University of California at Berkeley, established by Stanley in 1948. This nomination was carefully evaluated by Tiselius, who knew the TMV field well from his 1946 evaluation of Stanley. During the ten years that had passed Tiselius had become a Nobel Laureate himself, which gave additional prestige to his pronouncements as a member of the committee. He carefully reviewed the recent work nominated for the prize and also further publications in the field. In particular, and on his own initiative, he highlighted the work developed by Gerhardt Schramm and colleagues in Tübingen, Germany. He noted that their new data indicated that the nucleic acid by itself may be infectious and that this may shed new light on the work by Fraenkel-Conrat and Williams proposed for the prize. The infectious activity of the nucleic acid was not blocked by TMV-specific antibodies, but it was destroyed by enzymes destroying RNA. He wrote: "It may be possible that the ribonucleic acid plays the dominating role for the (infectious) activity and that the protein assists as an accelerator, a specificity-determining factor or only to stabilize the apparently rather instable nucleic acid component." After this visionary statement he concluded that this important research field and in particular Schramm's work should be followed carefully in the future. In later years, there were no more nominations of the potential candidates in this field for a prize in Chemistry. However the spectacular new findings were noticed in a later nomination for a prize in Physiology or Medicine.

In 1959 Edwin D. Kilbourne, Mount Sinai Medical School, New York, a forthcoming statesman of virology, made a comprehensive proposal which included Schramm and Fraenkel-Conrat. This proposal was discussed as a part of an impressively thorough, lucid and comprehensive review (10 out of 28 pages) by Georg Klein. He noted that it is now accepted that all viruses contain nucleic acid and that this can be either deoxyribonucleic acid (DNA) or RNA and that the question was if both these kinds of nucleic acids may carry genetic information. Some recent work with bacteriophages, to be further discussed below, indicated that DNA represented their genes. As remarked already by

Edwin D. Kilbourne, emeritus professor at New York Medical College. [Private photo.]

Georg Klein, professor of tumor biology at the Karolinska Institute. [Private photo.]

Tiselius in 1956 in his review, Schramm and Fraenkel-Conrat had gathered evidence that TMV RNA was infectious. Klein reviewed their experiments in detail. The approaches they used differed somewhat. Fraenkel-Conrat and collaborators[27] chemically separated the virus protein and the nucleic acid by use of a detergent and then let them reassemble again. In very clever experiments they used two different strains of TMV and let the RNA from one strain hybridize with the protein from the other strain. Such mixed particles could be inhibited by antibodies specific for the strain that provided the protein, whereas the particles produced after infection carried the properties of the virus that provided the RNA. This observation strongly indicated that the function of the protein shell was to protect the genes of the virus.

Schramm and his co-worker, the physicist Alfred Gierer, took a more direct approach[28]. They developed a method of isolating free high molecular weight RNA from the virus. This RNA turned out to carry the infectious activity. Klein considered the contributions prize-worthy because they opened a new field for studies of RNA genetics and they had significance for advances in our understanding of animal viruses. However, in the absence of a nomination of Gierer he refrained from arguing for a prize for the year concerned. Kilbourne's nomination of Schramm and Fraenkel-Conrat for "… the demonstration that nucleic acids are primarily responsible for the property of virus infectivity and the conveying of genetic information to the infected host cell …" also included the name Hershey to which Klein devoted the major part of his review. Hershey did his work with DNA-containing bacteriophages, which is the theme of the next section.

Bacteriophage replication and the studies of genes

If Stanley lacked capacity as a scholar or intellectual, this was not the case with Max Delbrück. He reminisced in his Nobel lecture entitled *A physicist's renewed look at biology—twenty years later*[29] about how he came to be involved in genetic studies and in the use of bacteriophages. The story had already been well told[8], and was briefly revisited in the lecture. It started out by recalling the continual discussion concerning the differences between living and non-living things and refered to Niels Bohr's famous lecture on life and light[30]. Although the latter article argued for some unique extra-molecular qualities of live materials it did inspire, together in fact with Stanley's 1935 finding of crystallized TMV, a write-up in 1937 of a preliminary note entitled *Riddle of life*. This note, co-authored by the Russian biologist and geneticist Nikolai Timofeeff-Ressovsky and the German physicist K. G. Zimmer, was attached as a memo to Delbrück's lecture. The conclusion of this text was that viruses should be considered as molecules, that the replication in the cell could be viewed as an autonomous accomplishment of the virus and that therefore one can "look upon the replication of viruses as a particular form of a primitive replication of genes."

Delbrück's Nobel lecture did not present any of the highlights of the findings he had made during his long career. There was in fact a problem with his Nobel Prize. The problem was that according to Nobel's will a prize in Physiology or Medicine can only be given for a *discovery*. Delbrück had inspired the general influential development of the phage field by his intellect and initiatives, but what paradigmatic discovery had he made? It is possible that Delbrück was aware of the particular requirement for the prize, since it is said that he first declined to accept the recognition, and then, after having been persuaded to do this, changed his mind. Fortunately, in the end, there came to be a triumph of the triumvirate and he was included together with Hershey and Luria in the award.

The bacteriophage field started in 1915–1917 with the findings by Frederick Twort and Félix d'Herelle and their contributions to the field have been reviewed[31]. Both of them were nominated repeatedly for a Nobel Prize in Physiology or Medicine. Twort, who had priority in the discovery, was nominated 11 times, and d'Herelle, who was more of an opportunist and better at advertising his contributions, 28 times. In 1926, the committee decided unanimously that the reserved 1925 prize should be awarded to d'Herelle for his discovery of bacteriophages and that the 1926 prize should be reserved. However, the faculty rebuffed the committee and sent the proposal back inquiring whether a proper evaluation of priorities had been made. The committee then decided that no prize should be given for 1925 and that the prize money for 1926 should be reserved. Eventually neither d'Herelle nor Twort received a prize.

The principal contributions of the phage field to the understanding of viruses were many. Instead of focusing primarily on the chemistry of the extracellular particles like in studies of TMV, these studies focused on the interaction between phages and their hosts, the bacteria.

Delbrück received a Rockefeller Foundation fellowship in 1937 which made him visit a number of laboratories in search of a project. One of them was the Nobel Laureate Thomas Morgan's laboratory at the California Institute of Technology. There he met the chemist Emory Ellis, who was working with phages in the context of cancer research. Delbrück saw the potential of working with phages for genetic purposes. They grow rapidly under controlled conditions and they could be quantified in the form of plaques, which could also have different appearances. Together with Ellis he further examined in 1939 the one-step growth curve of bacteriophages. This was the beginning of studies through the coming two decades of phage-bacteria interactions. Throughout these studies many findings of great importance were made.

There were intense controversies during the 1940s between Northrop and Delbrück[25], as already briefly mentioned, concerning the events in the infected cells. Northrop mentioned in his Nobel lecture that he had isolated a phage nucleoprotein and anticipated that it would crystallize like Stanley's TMV particles. Although he never achieved this, he continued to argue that viruses were produced in the infected cells by some autocatalytic process involving proteins and their precursors, as he had found with adaptive enzymes. It might be added that Stanley originally endorsed this idea, but changed his mind in 1937. In the end, of course, Delbrück was proved right in his interpretation of the virus as an independent genetic system which could reprogram bacteria to produce new virus particles.

Developments in the understanding of the nature of viruses by studies of bacteriophages were reviewed in 1955 in Malmgren's evaluation of Cohen, Delbrück, Hershey and Luria. In this comprehensive evaluation it was first noted that phages are now accepted as true viruses and that studies of them throughout the 1940s and 1950s have provided the most important insights into events occurring during the interaction between viruses and cells. The most studied bacterial viruses, the T-even phages of the common gut bacterium *Eschericia coli*, had been shown by the electron microscope to have a tadpole-

From left: Max Delbrück (1906–1981), Alfred D. Hershey (1908–1997) and Salvador E. Luria (1912–1991), recipients of the Nobel Prize in Physiology or Medicine 1969. [From *Les Prix Nobel 1969.*]

like shape and later demonstrated to contain DNA as a major component (page 226). Malmgren repeatedly returned to the critical experiment made by Alfred Hershey and Martha Chase, which was published in 1952[32] (Chapter 7). They had labelled the nucleic acid and the proteins with separate radioactive isotopes and demonstrated that it was essentially only the former that entered the bacteria. The review further went on to describe the important *eclipse* phase in the growth cycle of the virus that clearly distinguished it from bacterial or other cellular replication. This absence of infectious virus during the initial phase of its replication had already been observed in 1937 by Eugene and Elisabeth Wollman.

Various aspects of the destruction of bacteria, *lysis,* and of viral *interference* were also presented, but a discussion of the phenomenon of *lysogeny* was deferred. The evaluation further examined the studies of virus- and bacteria-specific protein and nucleic acid metabolism during infection. It then analyzed the two different important genetic phenomena, i.e. *transduction,* demonstrated in interactions between the phage genome and the bacterial genome, and *recombination,* which occurs between two viral genomes infecting the same bacterium. Plaque-type mutations of different kinds were used in the latter studies. Malmgren concluded that these studies of the virus-cell interactions were very important for the development of the field of virology at large. Without highlighting any particular discovery he recommended that all the candidates he had examined should be considered as prize-worthy. The committee in 1955 accepted that the candidates were worthy of a prize but delayed taking a position on the proposal to award it to all four of them, since a prize can only be given to a maximum of three people.

The 1959 nomination by Kilbourne also included, as mentioned, Hershey, who was analyzed in depth by Klein. He praised the Hershey and Chase experiment documenting the separate function of the phage protein coat and its DNA. A thorough background to what prompted the experiment was given and the consequences of the experiment were discussed extensively. It was now understood that a virus infection can be considered as a "parasitism at the genetic level" to use Luria's expression. Because of the genetic map already established by recombination experiments with the selected phages the possibilities for future advances were considerable. Klein concluded that the phage scientists have now come "... much closer to a chemical definition of the genetic material than in any other field. Heredity has now become a code construction problem." (Chapter 7.)

At the beginning of Kilbourne's nomination he mentioned in passing another important field of advances. This was the progress in formulating concepts of latent viruses, lysogeny and cancer pioneered by Lwoff. The basis for the interference phenomenon, eventually called lysogeny, was finally resolved by his elegant studies, in which the term *prophage* was introduced. This discovery was evaluated in depth in 1957 by Gard and it was revisited in Lwoff's scholarly and philosophical 1965 Nobel lecture entitled *Interaction among virus, cell and organism*[33]. The phenomenon of lysogeny, in which bacteriophages remained through generations of bacterial division without destructive consequences for the cells, was observed early.

Throughout the 1920s and 1930s it remained unexplained although Burnet proposed that it might be due to an absence of receptors for the virus. It was the Wollmans who, using the tools developed by Delbrück and Luria and others during the 1940s, first demonstrated the hereditary nature of the persistent infection. Lwoff then started an in-depth analysis of the phenomenon in 1949 and obtained a number of spectacular results. One of them was that the prophage could be activated by irradiation, leading to the death of the infected cell. It was furthermore found that prophage DNA was associated with the bacterial chromosome. This was demonstrated by use of the sexual mechanisms discovered in E. coli. This association could explain transduction phenomena. Gard did not hesitate to recommend that Lwoff should be considered as prize-worthy for his discoveries and this was also the conclusion of the Nobel Committee.

Gard also discussed briefly the possible relevance of the discovery of the lysogeny phenomenon for the understanding of animal virus infections. It was mentioned that herpes simplex virus infections in man are inducible, but generally there was some reluctance to give a major emphasis to the usefulness of insight into phage replication for understanding animal virus biology since "the principal difference between phages and most animal viruses is that the nucleic acid component in the one case is DNA and in the other case RNA." This remark should be seen in the perspective of the incomplete knowledge at the time of the presence of nucleic acids in different kinds of animal viruses. In the early 1960s we were purifying virus particles by gradient centrifugations and the type of nucleic acid they contained was examined by labelling with selective radioactively labeled nucleotides[34].

Studies of the transduction phenomenon used by Lwoff in his studies of prophages were pioneered by Lederberg (Chapter 6). In this context it deserves to cite his Nobel lecture[35]. This was entitled *A view of genetics*. It gave a grand survey of the field at the time. In a paragraph on *Genes and viruses,* Lederberg discussed the role of prophages in studies of bacterial genetics. Transduction was interpreted as a phenomenon when a phage in some way carries with it bacterial genes associated with its genome. There is also a short paragraph on *Viruses versus genes* which finishes "Does the virus have a unique element of structure, either chemical or physical, so far undetected? Or does it instruct its own preferential synthesis by a code of supporting enzymes?" In the final paragraph on *The creation of life* it became clear that DNA was then appreciated by Lederberg to be the fundamental information storage molecule of life.

Animal viruses and the understanding of the nature of viruses

Starting from the turn of the previous century ultrafiltrable agents causing disease in animals and man were demonstrated in increasing numbers. As summarized by Rivers[36], towards the end of the 1920s the viruses were defined by three negatives: they were not retained by filters withholding bacteria, they could not be seen in the light microscope and they could not replicate on artificial substrates that allowed the growth of different disease-causing bacteria. There were many loose speculations about the possible "organismal" nature of these minute infectious agents. At about the same time the relatively large particles of some viruses, like pox-viruses, were for the first time observed by dark field illumination and UV microscopy.

The animal viruses could only be studied by infection of the natural host. In the case of human viruses monkeys were generally used. The adaptation of different human viruses to small rodents and later to embryonated hen's eggs originally described by Ernest W. Goodpasture and further developed by Burnet, were important steps for studying them. Goodpasture was nominated seven times for a Nobel Prize in Physiology or Medicine and the four nominations in 1949, two of which included Burnet, were for the cultivation technique. In preliminary evaluations this technique was concluded to be important for the development of the field but it was never considered as a discovery worthy of a prize.

In 1958 there was a nomination of Richard E. Shope, Christopher H. Andrews and Wilson Smith for their discovery of the infectious agent causing influenza. This proposal was subjected to a thorough review by Gard. He presented the cumbersome search for the causative agent of this disease since the late 19th century. It took a long time before it was realized that the Gram-negative bacterium isolated by Richard F. J. Pfeiffer in 1893 and named *haemophilus influenzae* was *not* the cause of the disease. The pioneering work was done by Shope in studies of swine influenza. He identified an ultrafiltrable agent, but the interpretation of its etiological role was complicated by the need for additional factors, like bacteria, to give a full-blown disease in the animals.

Pandemic influenza in man is known throughout recent human history. Widespread occurrence of disease was recorded in the late 1840s, 1889 and the well-known and highly lethal "Spanish flu" in 1918–1919. Still, it took until 1933 before the virus that caused disease in man was eventually identified. This was done at the Medical Research Council, Mill Hill, UK. The group including Andrews and Smith was led by Patrick P. Laidlaw. He had earlier shown that the virus causing canine distemper, a close relative to measles virus in man, could infect ferrets. An attempt was therefore made to infect the same kind of animals with throat swabs from patients with clinical influenza passed through filters retaining bacteria. The material caused a biphasic fever in the animals and that it was the *bona fide* human influenza virus that had been isolated became apparent by an accident. One of the diseased ferrets had sneezed in Smith's face. Within a few days he developed a typical influenza and virus could be recovered from him. This virus was labeled by his initials and the WS strain came to represent the prototype of what later was called influenza A viruses. Influenza B viruses, a separate group was identified a year later. Andrews and collaborators made many pioneering studies of the influenza virus they had isolated. Success in growing the virus in the embryonated hen's egg in 1935 facilitated these. The size of the virus was determined and specific immunological reactions were identified.

In his summary Gard praised the discovery of the nature of the influenza viruses but in spite of this he did not recommend a prize. The fact that the leader of the group, Laidlaw, was no longer alive at the time of the proposal played an important role in this conclusion. Gard discussed a possible prize for Shope for his attempts to understand the persistence of viruses by some "masking" mechanism, a problem he had pursued further in studies of rabbit papillomas and fibromas. However, Gard noted that the critical advances in our understanding of viral persistence had been made in studies of phages, and that instead Lwoff and Lederberg should be considered as strong candidates for a prize for their important contributions.

The first experimental attenuated virus vaccine was developed by Theiler by passaging of yellow fever virus in chicken embryo cultures[16] (Chapter 4) and the first inactivated influenza vaccine was developed a decade later in the mid 1940s. Stanley, switching from work with plant viruses to those infecting man, got involved in the production and characterization of influenza virus particles at that time. No new findings of major importance were made and no crystals were obtained in his studies. The first genetic studies of animal viruses were made by examination of influenza virus pocks on the chorioallantoic membrane of embryonated hen's eggs. In addition the British virologist Leslie Hoyle used the egg system to characterize the growth curve of influenza virus. Importantly, he also identified the eclipse phase in replication of animal viruses. As summarized by Hoyle[37] it was not until towards the end of the 1940s that scientists studying animal viruses eventually started to discard the belief that viruses were rudimentary midget bacteria replicating by some kind of binary fission.

The opportunities to do biochemical work with animal viruses remained very poor throughout the 1940s with the exception of a few viruses that could be grown in large quantities in eggs. In spite of this Svedberg and Tiselius were encouraged by Stanley's success in crystallizing TMV to attempt a purification of the medically important polio virus. They raised money from a Swedish foundation in the early 1940s to set up a virus biochemistry laboratory at Uppsala University. Gard was appointed to lead the work in this laboratory and this is where he worked on his PhD thesis. Originally the intention was to use brains from polio-infected monkeys as a starting material, but because of the World War no animals of this kind were available. Instead Gard tried to purify virus particles from the brains of tens of thousands of mice inoculated with a polio-like virus. Although a considerable enrichment of the virus was achieved the true virus particles could never be identified by electron microscopy. A premature conclusion was reached that the virus showed similarities to TMV in terms of its nucleotide component and its molecular form. It was not until 1955 that poliovirus could be purified in sufficient quantities from tissue-culture-grown material to allow crystallization. This was achieved in the virus laboratory at the University of California at Berkeley.

Following the demonstration of the structure of TMV by electron microscopy the morphology of some other larger viruses of various origin was shown during the 1940s. It became apparent that viruses could show very different kinds of structures. Besides the rod-shaped TMV there were spherical plant viruses, the tadpole-like T-even phages, the large ellipsoid poxviruses, the spherical influenza virus, etc. Caspersson summarized the state of affairs in some detail in his 1946 evaluation of Stanley. He mentioned that more than 20 viruses had been (semi-)purified and that all of them contain nucleotides representing either RNA or DNA. Various chemical treatments of viruses, mostly employing TMV, could cause inactivation of the virus infectivity, but replication of surviving virus always led to production of virus similar, by the crude criteria employed, to the parental virus. However, the properties of the virus could be changed by X-ray or radioactive irradiation, which Caspersson concluded gives strength to the proposed analogy between a virus and a gene. Interestingly Caspersson discussed the interaction between the nucleic acid and the protein and came to the conclusion that they must be intimately connected. His thinking is in line with the prevailing thoughts of the time, that the nucleic acid served some packing function keeping the particles together. Finally he mentioned the variability in size and form of virus particles and their variable chemical composition. He states: "This is of importance as examples of the *unfounded conclusion* drawn by many authors, to regard viruses as giant molecules." It is not surprising that Stanley did not receive a prize in Physiology or Medicine.

An effective biochemical approach to studies of most animal viruses was not available until useful monolayer tissue culture techniques developed in the early 1950s. This kind of culture also facilitated the quantification of animal viruses using the plaque technique introduced by Renato Dulbecco. The prize awarded to Enders, Weller and Robbins in 1954 highlighted this breakthrough in the development of tissue cultures[17] (Chapter 5). Of course it was of paramount importance that they could demonstrate that poliovirus could replicate not only in neuronal cells but in essentially any kind of human or even other primate cells. It was this observation that paved the way for development of the successful vaccines against the disease. Still it should be noted that Gard concluded his address to the laureates at the prize ceremony[38] in the following way, "By giving the virologists a practical method for the isolation and study of viruses you relieved them of a handicap, burdening them from the birth of their science and placed them for the first time on an even footing with other microbe hunters." The number of different viruses isolated during the 1950s proved that Gard was right in his visionary statement.

The final synthesis of the virus concept

There were four major hurdles to overcome before a comprehensive interpretation could be made of the nature of viruses. These were:

1) Molecules engaged in life processes do not have unique properties — they are simply different kinds of organic substances.
2) Viruses are not a kind of mini-bacteria — they have their own mode of replication with an extracellular passive transport form and an intracellular piecemeal production and assembly process depending to a major extent on cellular metabolism.
3) Viruses can infect bacteria, plants and animals and the virus particles can have many different sizes, shapes and chemical composition — yet their general properties allow them to be grouped together.
4) Viruses contain genes and these are *not* represented by proteins, but by nucleic acids — in the case of viruses it can be either DNA or RNA.

To this list could be added that the genome of some viruses or copies of them can become a part of the cellular genetic material. Such an association can take many forms and allows the virus to persist, explaining various kinds of interference phenomena. It can also explain how viruses can influence the genetic functions of cells.

The old concept of *vitalism* has been very long-lived. Reinforced by different cultural concepts of religious origin it still lingers on in the minds of many people in present-day societies. The idea is that molecules involved in the processes of life must have some special qualities, possibly not even accessible by scientific techniques. The original interpretation of Stanley's crystallization of TMV was that there were dead and live proteins. Sedimentation analyses showed that the protein particle, interpreted to represent the virus, was large. Only later was it shown by British crystallographers that the particle was composed of repetitive units and a full understanding of its three-dimensional structure would have to wait for improvements in crystallographic techniques[15]. Even though Stanley's finding stimulated a lot of research, his argument that the virus was a giant protein blocked further development. The presence of RNA as demonstrated by

Bawden and Pirie did not stimulate the formulating of new concepts regarding the nature of viruses.

Initially it was logical to think about viruses as some kind of micro-organism that was parasitic in cells. In time it was understood that the virus particle had a structure, too simple to be capable of managing its own multiplication. Viruses appeared to have the qualities of life associated with replication and mutation, but in order to manage energy-dependent metabolism they needed cells[39]. The mode of replication became understood from studies of phages during the 1940s and influenza virus later during the same decade. In order to fully grasp the nature of the cell-destructive — lytic — cycle the basis for other kinds of virus-cell interactions like homologous and heterologous interference and latency had to be clarified. Later on the term "lytic" had to be further qualified, when it was found that in some cases new viruses are produced progressively over a longer period of time instead of being released simultaneously by disruption of the infected cell.

It was natural that it took time to bring together ultrafiltrable infectious agents replicating in such different kinds of cells as those from bacteria, plants and animals. Furthermore they had to be distinguished from rickettsia, which turned out to be cellular parasites of bacterial origin. As more was learnt it was, however, finally understood that the many different kinds of viruses in the extra-cellular phase of their life cycle were simply various forms of packages of genes.

As in the general field of molecular biology the major hurdle that had to be overcome to understand the nature of viruses and comprehend their replication was the eventual dramatic insight that nucleic acids and *not* proteins represented the genetic material. These developments will be discussed in more depth in Chapter 7, but it can be noted that Dubos argued[40] that as early as 1943, Burnet, when visiting Avery's laboratory at the Rockefeller Institute, accepted that DNA represented purified genes. However, it is not clear from the writings of this visionary virologist that he believed that nucleic acid also possibly represent the genes in viruses. Still, in 1955, in his popular book *Viruses and Man*[41], Burnet stated "… a gene is probably a *pattern* carried mutually by a nucleic acid molecule and a protein molecule, but the nature of the pattern cannot be defined in chemical terms." Furthermore in the same book he also stated apropos Stanley's Nobel Prize "This work was in the very best tradition of modern biochemistry — and yet there is more than a suspicion that it has not got us very far." In defence of Stanley it should be mentioned that Burnet might have been more interested in what viruses do than what they were[5]. In the context of Burnet's statement about Stanley it can be noted that surprisingly still in 1970[42] the chairman of the Nobel Committee for Chemistry during the years 1944–1964, Arne F. Westgren stated: "Research on this subject deals with the borderline between living and dead matter. The progress made in this field, particularly owing to Stanley's work, is most impressive and has attracted much interest among scientists."

Since the mid 1930s it had been understood that viruses were nucleoproteins. As more viruses were examined the presence of nucleic acid was consistently identified. However the crystallization of the TMV particles, stimulating to virus research as it was, meant that there was an intensified conceptual focus on their protein part. The question became how this "live" protein could replicate and the multiple hypotheses formulated led into many blind alleys of research. This focusing on protein probably overshadowed any serious consideration of a major role for the nucleic acid. The conceptual deductions from the work with TMV probably also came to taint the interpretation of findings made in studies of phages. It is said that the outcome of the Hershey–Chase experiment in 1952 came as a surprise even to Hershey himself. And it is notable that he does not mention this critical experiment in his Nobel lecture[43].

The demonstration in 1956 by A. Gierer and G. Schramm and by H. Fraenkel-Conrat[27,28] that TMV RNA is infectious and could reassemble with its coat protein into intact particles can be said to have closed the case. Even though their contribution was very important to virology and also to biology in general, it never came to be recognized by a Nobel Prize. Within a few years after their discovery, RNA isolated from polio virus and other animal viruses were demonstrated to be infectious and in 1961 DNA isolated from bacteriophage φX174 was also documented to be capable of infecting cells. Virus researchers could have become instrumental at a relatively earlier time to force the paradigmatic shift from a focus on proteins to nucleic acids as representing the genetic material, but for various reasons that never came to be.

A counterfactual history of virology

The particular features of the two dominating kinds of viruses used in the early search for the nature of viruses, TMV and the T-even phages, naturally influenced the steps of development of new knowledge. It could be of interest as a "thought" experiment to reflect on what course the advance of the virus concept would have taken if TMV had been a virus of a different nature. If it had had an icosahedral instead of a helical symmetry it might have been easier to find a homogenously sedimenting population of particles, but extended crystallographic studies would not have allowed a definition of the detailed three-dimensional location of its RNA. If TMV in addition to being spherical had had its genome divided into separate populations of particles, as in some other plant viruses, additional problems in the analyses would have arisen.

Another possibility is that TMV would have been an enveloped virus, like the plant rhabdoviruses and bunyaviruses later discovered. Then no virus crystals would have been produced in the laboratory and the discussion of the dead versus live protein concept might not have been as prevalent. Although such infectious agents have not as yet been encountered in plants, one could hypothesize about TMV being a virus with a very large genome, as in herpes—or poxviruses, or why not even like the mimivirus recently discovered in an amoeba[44]. The latter virus has a genome that is 1.2-Mb large directing the synthesis of 1,262 protein-coding genes. This is more than twice as much as the number of genes required for the autonomous replication of the simplest form of cellular life, a mycoplasma. Genes that code for enzymes and factors involved in translation have now been found in viruses. Had TMV been this kind of virus, a distinct borderline between virus and organismal microorganisms would have been more difficult to draw.

We can also go to the other end of the scale and speculate on TMV being a viroid. These predominately plant pathogens consist solely of a short stretch of highly complementary circular single-stranded RNA, which does not code for any known protein. If TMV had been this kind of infectious agent and some chemist had managed to purify it as early as the 1930s or 1940s, the discussion would not have focused on the mind-boggling replication of an infectious protein, but instead on an infectious RNA, a kind of molecule at the time considered only to have support functions. Finally if TMV had been a prion the discussion would have been in reverse (Chapter 8). Then Stanley would have been right in his argumentation for an infectious protein!

Virology 50 years later

There have been impressive developments in the field of virology during the last 50 years. To a large extent virology has become a molecular biological science. Many different genomic strategies used by viruses have been elucidated. The genomes of some simpler viruses have been synthesized and shown to be infectious. Impressive advances have been made in the development of vaccines and also other means to prevent or treat virus infections. But to what extent have we widened our perspective on the ubiquitous viruses to concern not only their role as pathogenic agents?

Like bacteria, viruses were discovered in a context where they cause disease processes in plants and in animals. Due to our anthropocentric focus we naturally came to search for and learn to know the characteristics of viruses that cause disease in man and in the domesticated animals and plants that we exploit for our living. The discoveries made reinforced our perception that viruses are the cause of disease, or as expressed by Medawar, "a package of bad news."

Extended studies of bacteria revealed that only special kinds were the cause of disease and that the overwhelming numbers of them in nature were "harmless." Bacteria identified to date represent about 6,000 different species, which can be categorized into two large domains called Bacteria and Archaea. The prokaryotes together with microbial Eucarya (protists, fungi and algae) constitute the totality of the invisible world of microorganisms. They represent a larger biomass on Earth than all plants and animals together. However, only a very small fraction of all bacteria represented in our world have been identified. A limitation to our studies of the existence of different bacterial species is that only a small fraction of them, speculatively less than one percent, can be cultivated in the laboratory. Only in the last few years has it been possible also to characterize non-cultivatable bacteria. Their complete genomes can be identified by a *metagenomic* analysis, combining new nucleotides sequencing technologies and advanced bioinformatics. In the near future we will learn to know many hundred thousands of new bacterial species. It serves to remind us that our bodies harbor more than ten times more bacterial cells than the ten trillions of cells that constitute our human organism. We are a walking society. The diversity and function of this flora, the *microbiome*, is now examined by metagenomic analysis.

In the discussions of biodiversity in our world, the wide representation of microorganisms has generally been neglected. But what is currently even more overlooked is the presence of viruses, the most ubiquitous of all evolutionary entities. The genes of viruses successfully survive in nature and hence represent an important part of the total evolutionary toolkit. Most importantly wherever there are cells there are viruses — *ubi cellula, ibi virus*. It can be projected that for each species of micro-organism, plant and animal there are a minimum of 10–100 viruses. More than 1,000 different viruses have been documented to infect man, a single species. And then we have only counted the viruses that infect our nucleated cells and not those that infect all of our companion bacterial cells. In the future it will be possible not only to perform a microbiome analysis of all our bacteria but also to make a *virobiome* examination of the totality of all viruses in our body.

One consequence of the ubiquitous representation of viruses is that the tree or, rather, considering the dominance of bacteria, the web of life is surrounded by a huge cloud of viruses! This is not the way it is presented today. Thus if it be speculated that there are at least some 100 million species of animals, plants and microorganisms then there would be as many as a billion or more different species of viruses! Studies have been initiated of viruses associated with all different forms of life and not necessarily only as a cause of damage or disease. This inventory is only in its initial stages but a lot of interesting data are now beginning to be presented. There are, for example, of the order of 10 million virus particles per ml of surface seawater in our oceans. It is therefore justified to expand the discussion of the role of viruses, way beyond their importance as pathogens. The function they have for the total ecological balance in nature and more generally for the evolution of life[45] needs to be seriously investigated and this may be even more important than the role viruses play as pathogens. This, however, is a theme for a more forward-looking discussion than the history of the emergence of the virus concept reviewed here.

Coda — Are viruses live or dead material?

Many answers have been given to this question over the years. It has been discussed in relationship to the two forms of viruses described above: the extracellular passive transport form and the intracellular events of virus-controlled synthesis of new particles. However, I would argue that the question is incorrectly posed. In order to answer the question one needs to find a resilient definition of what life is. In the search for extraterrestrial life the general definition used is that life is *a (self-sustaining) replicative chemical system subjected to Darwinian evolution.* Interestingly, a prerequisite is that the system can make mistakes. When discussing life on Earth various cellular forms of life are generally referred to. However, different cellular forms of life show a highly diverse complexity and a wide range of dependence on their environment. Already in 1973 the famous biologist Theodesius Dobzhansky stated: "*Nothing in biology makes sense except in the light of evolution.*" In line with this statement and in the absence of a coherent general definition of cellular life an alternative question to ask is: "*Is biological entity X, e.g. a virus, participating in Darwinian evolution?*"

Whatever form of replicating system that exists spontaneously in Nature by definition has a capacity for survival under the given conditions. Descartes' famous dictum *Cogito ergo sum* — I think therefore I exist — might be paraphrased to *Existo, ergo pars evolutionis sum* — I exist, therefore I am a part of evolution. If we therefore ask the critical and appropriate question "Do viruses participate in evolution?" the answer we get is a resounding "Yes." Viruses, as well as the subviral biological entities mentioned above, are indeed active in evolution. In fact there is considerable evidence that viruses may even be the prime movers in the development of new forms of life by evolution. This perspective accentuates the need to expand studies of the ubiquitous — and possibly generally harmless — viruses as proposed above. However, traditionally we have hitherto been concerned mostly with viruses that cause damage to ourselves or our cultures. The next chapter will concern one of the scourges of mankind, yellow fever, and the success in combating this virus disease through a discovery rewarded by a Nobel Prize.

Chapter 4

The Only Nobel Prize for a Virus Vaccine: Yellow Fever and Max Theiler

In 1951 Max Theiler of the Rockefeller Foundation received the Nobel Prize in Physiology or Medicine "for his discoveries concerning yellow fever and how to combat it." This was the first, and so far the only, Nobel Prize given for the development of a virus vaccine. Dr Theiler was a scientist with a very retiring disposition. He rarely traveled and published sparsely. He preferred to stay in his laboratory and there he finally managed, during the 1930s, to develop a safe and effective yellow fever vaccine. He succeeded where many had failed before him. For him the appearance in the limelight of the Nobel festivities in Stockholm during the week surrounding the prize award ceremony on December 10, 1951 was an exceptional event (see photo on page 98).

Yellow fever disease has caused life-threatening epidemics throughout the last 500 years of human civilization. In the first half of the 20th century it took extensive studies, with major sacrifices by those involved, before the nature of the disease could be clarified. Once its viral origin had been identified, its means of spreading in nature could be clarified, and it became possible to search for ways to prevent the disease. The concluding advance in these studies was Max Theiler's development of the 17D strain of attenuated virus, which could be used as a live vaccine to save the lives of many millions of people. There was no question that the introduction of this vaccine was "to the benefit of mankind" as is specified in Nobel's will, but how does Theiler's contribution compare with other advances leading to vaccines against viral diseases, introduced both earlier and later? Was it based on a true discovery?

Max Theiler (1899–1972) receiving the 1951 Nobel Prize in Physiology or Medicine from the hands of His Majesty King Gustaf VI Adolf. [Courtesy of the Karolinska Institute.]

This chapter will discuss how recently-released Nobel archives shed light on how the yellow fever vaccine field was evaluated in the late 1940s and how this led to a prize for Max Theiler. The Nobel archives remain closed to researchers for 50 years[1], as repeatedly mentioned. Three kinds of material are available for reviewing: the nominations submitted, the reviews made by professors at the Karolinska Institute and the recommendations of the Nobel Committee for Physiology or Medicine to the College of Teachers. Candidates considered to be prize-worthy are listed and the candidate(s) recommended for the prize in a certain year by the majority of the committee is/are presented. The committee was composed of three or five ordinary members, but it was enlarged a long time ago by the addition of at least an equal number of adjunct members. In the early 1950s the sole responsibility of the Karolinska Institute was to train physicians. Its College of Teachers, the faculty, encompassed some 35–40 tenured professors. The recommendation by the committee to the faculty is in the form of a decision, not a discussion, protocol. Thus, divergent arguments and opinions of individual members cannot be identified except in the form of non-motivated reservations to the protocol.

This chapter also serves to illustrate the continuously evolving mechanisms, managed by the collective efforts of different groups of scientists, by which the choice of a particular Nobel Prize recipient is made in a particular year. It is notable, in this regard, that the yellow fever vaccine has been singled out, as the only representative among a large group of very important medical compounds used for prophylactic interventions, to deserve recognition by a Nobel Prize. In order to understand this, the latter part of the chapter gives a brief review of the development of different vaccines and discusses the concept of "discovery", which was specified by Alfred Nobel as the one and only criterion of a prize in Physiology or Medicine.

The disease and the epidemics

"All the medicines in the pharmacopoeia — the doctors seemed determined to use them all — had not power enough to arrest disaster or erase the horrid scenes presented in these two first weeks of September. Terror and numb dismay overwhelmed people. The poorest in spirit abandoned themselves to primitive efforts at self-preservation, sometime forsaking the simplest decencies of human relationships. In the continual failure of the doctors even the bravest men lost hope. The city surrendered to a coarsening fear." This is the introduction to the chapter "Panic" in the colorful book by J. H. Powell, *Bring Out Your Dead. The Great Plague of Philadelphia in 1793.*[2] The 1793 epidemic in the temporary capital of the young United States of America paralyzed city functions, halted business and trade and caused a breakdown of societal functions. The epidemic, probably introduced by French refugees from Santo Domingo, ravaged the city for three months from the beginning of September until the first night of frost in November. Out of a total population of about 50,000 inhabitants, 17,000 fled the city and about 5,000 died. This was the first epidemic since 1762, but succeeding epidemics came soon afterwards, in 1797, 1798, and 1799. However, the 1793 epidemic was the worst and remains the one best analyzed. It was even hailed in a poem by Philip Freneau, written at the time:

Nature's poison here collected.
Water earth and air infected.
O, what a pity
Such a City
Was in such a place erected.

A decade later the disease came to play a major role in the history of the West Indies, and indirectly also of the US. In 1802 there was a revolt among the slaves on the sugar plantations of the French colony of Haiti, led by Toussaint l'Ouverture. The reason for this uprising was that slavery, which had been abolished after the French revolution, had been reintroduced by Napoleon Bonaparte. To suppress the revolt he sent some 30,000 of his best troops, but because of a rampant yellow fever epidemic, the expedition turned into a disaster. A large majority of the soldiers in the French army died — including the expedition's commander and Bonaparte's brother-in-law, Charles Leclerc — and although l'Ouverture was brought to France the mission failed. This failure had major geopolitical consequences. Napoleon backed off from his ambitions to rebuild France's New World empire and instead, in order to raise money for his wars in Europe, he decided to sell a much larger piece of land to the United States than originally was intended. The negotiators, James Monroe and Robert R. Livingston, sent to France by Thomas Jefferson, were on a mission to discuss a deal involving only New Orleans and its surroundings. The young US eventually bought 2,147,000 square kilometers (23% of the territory of present day US) for a total cost of 15 million dollars. Upon completion of the agreement Napoleon stated: "This accession of territory affirms forever the power of the US, and I have given England a maritime rival who sooner or later will humble her pride."

Another situation where yellow fever came to have a decisive influence on important developments in the Americas concerned the project to build a canal through the Panamanian isthmus. The Frenchman Ferdinand de Lesseps — organizer of the successful Suez Canal project — initiated this project in the 1880s, but after the death of about 22,000 workers, due to yellow fever and malaria infections, he had to give up. In 1904 the US took over the project. Four years earlier a US Army Yellow Fever Commission had been established to cope with the yellow fever problem in Cuba. During that time the US was engaged in the war with Spain that in 1902 led to the partial independence of Cuba.

Headed by the US army surgeon Walter Reed, the Commission had managed in heroic experiments to show that a mosquito vector, *Aedes aegypti*, spread yellow fever. This was a confirmation of a theory formulated some 20 years earlier by the Cuban doctor Carlos Finlay. Reed's collaborator Jesse Lazear was instrumental in designing and conducting the critical trials. He succumbed to yellow fever after having become experimentally infected by the mosquitoes he worked with. Some other infected human volunteers also died from the disease (only later during the studies were informed consent documents developed). Reed himself died unexpectedly from appendicitis in 1902, but his contribution was considered so important that he was nominated four times posthumously (1903–1905) for a Nobel Prize in Physiology or Medicine. Since a nominee has to be alive at the time the decision about a Nobel Prize is made, Reed's contribution was never evaluated.

Walter Reed (1851–1902), who demonstrated that yellow fever is caused by a virus. [Courtesy of the Office of History, US Army Medical Department.]

Based on the knowledge accumulated by Reed and his collaborators, the US Panama Canal project was pursued under conditions of extreme sanitation supervised by US Army surgeon William C. Gorgas. One of his collaborators was Ronald Ross from Britain, who in 1902 had received the second Nobel Prize in Physiology or Medicine for his discoveries from studies in India that malaria is also transmitted to man via a vector, a special group of mosquitoes, *Anopheles*. The effects of the intense measures introduced were dramatic. Both yellow fever and malaria were essentially eliminated from the region and the canal-building project could be brought to a conclusion. Gorgas' successful work with sanitation in connection with this project was so highly regarded that he was nominated six times in 1909–1919 for a Nobel Prize. However, the Nobel Committees never subjected his contribution to an investigation. It did not include any independent discovery.

The effective elimination of yellow fever from the Panamanian isthmus led to the general view that the disease might be eradicable. This view had to be modified as more was learned about it. The Rockefeller Foundation, established in 1913, came to play a decisive role in increasing understanding of the disease[3]. In 1915 a Yellow Fever Commission was established by the closely-related International Health Board, which also was Rockefeller-funded. The primary goal was to eliminate breeding places for *Aedes aegypti* in areas where the disease was prevalent. Successful interventions were made in Ecuador and later on in Brazil. These efforts were then expanded to include the other afflicted continent, Africa. The first West Africa Commission, established in 1920, was led by Gorgas, who unfortunately had a stroke in London on his way to Africa. Before his death King George V came to his bedside and bestowed a knighthood on him. Under the new leadership of General R. E. Noble the Commission initiated its work in West Africa and in

1925 laboratory headquarters were established. Many important discoveries came to be made in this laboratory, but it was also a place where major tragedies were to occur, as we shall see.

It was important for the emerging understanding of the epidemiology of yellow fever when field studies in the 1930s could demonstrate that the disease could circulate, in principle, under two different conditions. Its natural reservoir is monkeys, between which the infection spreads by various jungle-dwelling mosquitoes. The virus can occasionally be transmitted from infected monkeys to humans by use of a range of different vectors and result in individual or small clusters of cases. This is referred to as *jungle* or *sylvan* yellow fever. If, however, these occasional cases of yellow fever come into contact with larger populations of humans in urban areas, while the *Aedes aegypti* vector is present, a severe epidemic can develop. In this *urban* form of yellow fever the virus is effectively circulated from person to person.

The virus has its origin in Africa and spread from there during the 17th century to the New World, first to the Yucatan peninsula and the Caribbean islands on board ships, which carried both infected individuals and the critical vector. For hundreds of years the *Yellow Jack*, as the disease was called among seafarers, was a hazard for maritime ventures. The name yellow fever is derived from the jaundice that occurs late in the disease in some patients due to liver failure. The infection leads to major hemorrhages throughout the body and the bloody — "coffee ground" — vomit has given the disease its Spanish name, *vomito negro*.

Since its spread across the Atlantic, yellow fever has been in residence in both Africa and the Americas. There have been concerns that it could spread to other parts of the world, and severe epidemics have indeed developed in Europe, for example in Spain. However, epidemics have not been seen in Asia, perhaps because the virus and the vector could not survive the long sea voyage.

In the 1950s the accumulated results of the widespread use of the vaccine that Theiler had developed in the 1930s gave such confidence that the first chapter by Andrew J. Warren in a comprehensive book summarizing all the achievements of the Rockefeller Foundation until 1951[4] was entitled "Landmarks in the Conquest of Yellow Fever." The additional results up until today have shown that it is more appropriate to use the term "control" than "conquest." Because of the safety and efficacy of Theiler's vaccine virus[5] the control is highly effective, but there are still, even at the present time, flare-ups of sometimes rather extensive epidemics in the tropical belts of the Americas and Africa[6]. It is estimated that there are about 200,000 cases and 30,000 deaths every year in non-immunized populations. Much has been learned during the last 50 years about the complex interactions between different hosts, various vectors and strains of the virus[7]. There will always be a need for vaccination in the countries afflicted, since the jungle source of the virus can never be exterminated. Let us now go back in history and consider the conditions for control of the disease and the final successful development of the vaccine that has come to play such a critical role in this.

The virus and possibilities for vaccine development

Walter Reed and James Carroll, by the use of courageous collaborators and Army volunteers, had shown in 1902 that the agent causing yellow fever could pass through bacteria-proof filters[8]. This was the first human infectious agent to be shown to be *ultrafiltrable* (Chapter 3), but it took time before the scientific community was convinced of its virus nature.

In the 1920s a devoted and persuasive scientist of Japanese origin, Hideyo Noguchi, working at the Rockefeller Institute, had a marked influence on approaches to this problem. He believed, and managed to convince many others, including its director, Simon Flexner, that a bacterium, *Leptospira icteroides,* and not a virus was the cause of the disease. Both Noguchi and Flexner appear frequently in the files of nominations for the Nobel Prize in Physiology or Medicine. Noguchi was nominated no fewer than 24 times during 1913–1927, originally for his work on the spirochete causing syphilis, but later predominantly for his hypothesis that a spirochete was the causative agent of yellow fever. He seems to have had many supporters around the world[9] in spite of the fact that it was shown later that most of his experimental findings could not be reproduced. Flexner was mainly nominated for his work on serum treatment of epidemic meningitis and also for his studies of the aetiological agent of poliomyelitis. In 1914 the professor of pathology Carl Sundberg made a full investigation of both Flexner's and Noguchi's work and this was supplemented by another full investigation of both of them in 1915. Neither was considered prize-worthy. Ten years later Noguchi received as many as eight nominations, and the 1925 Nobel Committee had the professor of forensic medicine Gunnar Hedrén carry out a thorough new investigation. Fortunately neither Hedrén nor the committee was seduced by Noguchi's enthusiastic pushing of his hypothesis.

Stokes and collaborators at the Rockefeller Foundation laboratory in Nigeria managed in 1927 to show that monkeys could be infected with

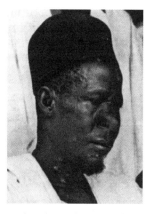

Asibi, the individual from whom the first yellow fever virus was isolated. [Photo from Reference 4.]

material from humans with yellow fever[10]. This was an important breakthrough. The viral nature of the agent could be conclusively confirmed by use of this animal system. The virus was named the Asibi strain from the name of the 28-year-old West African yellow fever survivor, who provided the blood sample. This virus strain came to play a central role in the eventual successful development of a vaccine. One of Stokes's collaborators, Johannes Bauer, brought the virus to the Rockefeller Foundation laboratories in New York.

When Noguchi heard that Stokes's experiments showed that the agent was a virus and not a spirochete he decided to join the Rockefeller laboratory in West Africa. By the time of his arrival in January 1928 Stokes had died four months earlier of yellow fever, and it did not take more than five months until Noguchi himself succumbed to the disease. The director of the laboratory W. A. Young performed the autopsy to verify the cause of Noguchi's death. Nine days later Young himself also died of yellow fever. These tragic sacrifices came to foster the image of yellow fever researchers as heroes among microbe hunters. In the 1934 Broadway production *Yellow Jack* Jimmy Stewart played a brave and enthusiastic researcher-guinea pig. Noguchi has been recognized as a hero in Japan until these days. Since 2004 his picture can be seen on 1,000 yen banknotes.

Throughout most of the 1920s all experimental yellow fever research supported by the International Health Board of the Rockefeller Foundation was developed in the laboratory it had built up in West Africa. However, during the same time the possibility of establishing an experimental laboratory affiliated with the Rockefeller Institute of Medical Research in New York was explored by its director Fredrick Russell. The powerful director of the Institute, Simon Flexner, opposed these plans because of the risks of laboratory infections. Therefore it was not until 1928 that a laboratory for experimental work with the virus was established on the campus of the Institute. This laboratory was headed by Wilbur A. Sawyer. It became the most important center for work with the virus in the world. The physical association of this laboratory and those of the Institute was probably of great value in these developments.

Conditions for experimenting with monkeys had been established and basic general properties of the virus could be analyzed. It turned out that it was a relatively small and somewhat unstable virus[11]. The presence of proteins stabilized its infectivity. In present-day classification the yellow fever virus is grouped together with more than 50 other viruses carried by arthropod vectors. The group is named "flavivirus," from Lat *flavus* meaning yellow, a collective name appropriately deriving from its most prominent member. Once it was understood that yellow fever could not be eradicated by hygienic measures it was clear that a vaccine had to be developed to control the disease. The only system available to produce virus in the late 1920s was the use of infected monkeys. It was not possible to use this kind of animal for production of a vaccine and for determination of protecting antibodies in sera, but the infected animals provided a valuable model of what might happen in human infections. The capacity of a certain virus strain to replicate in different organs could be experimentally evaluated.

There were a number of options for further progress in the development of a vaccine. The many different scientists engaged in this field of research took several approaches. In the first place a better system for practical production of virus for vaccines was needed. Could one use an alternative animal system, or could the crude forms of tissue cultures available at the time be used? Secondly, some method for attenuation of virus must be found. Repeated passage of virus under some kind of unnatural conditions was the approach at the time. Once this empirical probing had resulted in some virus progeny possibly worth characterizing, the next question was how this should be done.

How would one estimate the projected capacity of the virus to replicate in man, both from a quantitative — the total amount of virus produced — and qualitative — in which organs did the virus preferentially reproduce — perspective? The latter quality, which is usually referred to as tropism, can, in the case of yellow fever, be divided into two essentially separate properties. The virus can be *viscerotropic*, i.e. it can replicate in many internal organs generally combined with bleedings as seen in a typical case of yellow fever in man. The virus can also be *neurotropic*, i.e. it can replicate effectively in the brain and in the peripheral nervous system. Some virus strains show both these characteristics. The ideal live vaccine strain should display none of them, but still replicate well enough in the vaccinated individual to induce a durable immunity.

In the case of a live yellow fever vaccine there is no alternative to injecting the virus or depositing it by scarification of the skin, since this is the normal step of entry for this vector-borne agent. However, the dose of virus can be varied, and in the early phase of research attempts were made to reduce the intensity of replication of test vaccine preparations either by diluting them, by partially inactivating their infectivity (for example by heating), or by a simultaneous administration of a selected dose of immune serum.

Theiler's way through all these alternative experimental approaches for the development of an effective vaccine will be described below. But first let us consider his scientific career.

Max Theiler, the experimental scientist

When the news came that Max Theiler had been awarded the Nobel Prize in Physiology or Medicine in 1951, Dr Robert S. Morison, associate director of the Division of Medicine and Public Health at the Rockefeller Foundation got a call from William Laurence, science editor of *The New York Times* at the time. He lamented: "Theiler isn't in *Who's Who*, he isn't in *American Men of Science*. I know that you people like to soft-pedal your work, but this is ridiculous!" Regrettably I never met Dr Theiler personally, but Sven Gard (page 66), my mentor and predecessor in the chair of Virology at the Karolinska Institute, knew him well.

Gard had received a Rockefeller Foundation scholarship to carry out research for one year (1939) in the US, out of which ten months were to be spent in Dr Theiler's laboratory. This was a very productive visit. Gard worked primarily with Theiler's virus, a poliovirus-like agent that replicated in the brains of mice. His collaboration with Theiler resulted in two publications[12,13]. Gard did in fact also perform some work together with John Fox on the important issue of the stability of yellow fever virus using the technique of desiccation[14].

I visited the Foundation's laboratory at Rockefeller Institute for the first time in 1962, but the letter of introduction from Gard was to the epidemiologist of the Rockefeller Foundation, Jordi Casals, and not to Theiler. My visit included a meal in the classical lunchroom with its round tables that was then still in existence. This lunchroom in which iced tea — my first encounter with this beverage — was served with the meal is famous in the annals of the history of science as representing an environment that uniquely furthered creativity[15]. In a letter of January 1939 introducing Gard, Johannes Bauer wrote to the then director of the Institute, Herbert Gasser: "I hope that it will meet with your approval to have Dr Gard work with us at the Institute and that during his stay he will be granted the privilege of using the lunchroom and library." Nothing was said about access to laboratory facilities. Gasser was to be awarded the

Nobel Prize in Physiology or Medicine for 1944 in 1945 (Chapter 6), one of many to be honored in this way at the unique Rockefeller Institute.

Both from Gard and from other people who have met Theiler, like Dr Harriet Zuckerman, the author of a book on American Nobel Laureates[16], who interviewed him in October 1963[17], one hears the same description of him as an exceptionally modest, gentle and unassuming person. He never sought public attention and, surprisingly, never became a member of the US National Academy of Sciences. He did finally make it to *Who's Who*, however. His bibliography is relatively short, but apparently he managed, staying in his laboratory, to make the proper intuitive choices to find solutions to very important practical problems. So who was this virtually anonymous man, Max Theiler?

He was born in Pretoria, South Africa in 1899 to Swiss-born parents and spent the first 20 years of his life there. After leaving South Africa he never returned, but he retained his citizenship. He also later acquired US citizenship. His father was a famous professor of veterinary medicine, who did pioneering work on filterable agents and certainly inspired his son in his future career. After premedical training in South Africa, he left for England to do his medical training at St. Thomas' Hospital in London. In 1922 he became a Licentiate of the Royal College of Physicians and a member of the Royal College of Surgeons. In the same year he also received a Diploma of Tropical Medicine and Hygiene. He then went to the US and between 1922 and 1930 was first an assistant and later an instructor at the Department of Tropical Medicine at the Harvard Medical School in Boston.

The head of the department, Dr A. W. Sellards, had a particular interest in yellow fever. Following the success of Stokes *et al.*, he and his collaborators, working in Dakar, French West Africa, had also isolated the virus in monkeys[18]. This isolate, called the French strain, was the first experimental yellow fever virus brought to the US. Theiler worked on various tropical infections in the laboratory and made his first contribution to the yellow fever field when he showed that *Leptospira icteroides* was identical with *Leptospira icterohaemor-rhagiae*, the cause of Weil's disease[19]. This finding argued against the prevailing opinion of the time, originating from Noguchi's results, that this agent had anything to do with yellow fever. Soon thereafter the issue was settled, as has already been mentioned, when the Rockefeller Foundation group led by Bauer showed by their monkey experiments that the causative agent was ultrafiltrable. Theiler also performed some preliminary comparative immunological studies of yellow fever virus from West Africa and South America[20]. His different findings might have been used in a PhD thesis, but Theiler was never formally examined for this degree.

The next important development in his career was when he searched for an alternative to the expensive and cumbersome use of monkeys for studies of the virus. He was successful in propagating the French strain of virus in the brains of mice[21,22]. This finding offered a radically new approach to yellow fever research. Because of this contribution Theiler was heartily welcomed to the Rockefeller Foundation when in 1930 he applied for a post in its International Health Division (the name of the International Health Board since its merger with the Rockefeller Foundation in 1928). This association with the leading laboratory in the world for yellow fever research was an important step in Theiler's career. It was the right place for the development of his unassuming creative capacity.

In the interview with Dr Zuckerman[17] he emphasized the congenial and informal atmosphere of the laboratory. He clearly preferred that to a very competitive environment. He stayed with the Foundation throughout the rest of his professional life, retiring in 1964. For the last 13 years he was the director of the virus laboratory. At the time of his retirement the Foundation's laboratories moved to Yale University, New Haven, Connecticut, where Theiler had some rewarding years as an emeritus professor and became very much liked by students in the new university setting. He was remembered as a most engaging and approachable person, always ready to study a proposal and to provide friendly criticism, and always interested in watching and discussing the progress and outcome of experiments.

Outside the laboratory his interests were baseball and reading. He liked to rise early to do his reading in philosophy and history, for example, *A Study of History*, the 12-volume *magnum opus* by Arnold J. Toynbee. His concern for his employees is apparent from the following memorandum of July 25, 1933, to his director to recommend an annual salary raise from $140 to $150 to Mr T. W. Norton: "Mr Norton was the first of the laboratory technicians. He joined the staff in 1928. Since then he has had yellow fever and Rift Valley disease as the result of laboratory infections. Both attacks were moderately severe. While convalescent from Rift Valley disease, he had appendicitis and contracted a heavy debt to surgeon and hospital. He has become an expert in working with mice and makes inoculations for me. He is very dependable."

Theiler received the Chalmers Medal in 1939 from the Royal Society of Tropical Medicine and Hygiene, England; the Flattery Medal, awarded by the Harvard University in 1945; and the Lasker Award, given by the Lasker Foundation, in 1949. He died in 1972.

The first deliberations by the Nobel Committee

In order to be considered for a Nobel Prize for a certain year an individual has to be nominated before January 31 the same year. Only specially-designated individuals or persons representing organizations invited by a Nobel Committee to submit nominations are eligible to propose candidates[23] (page 24). Max Theiler was proposed for the Nobel Prize in Physiology or Medicine for the first time in 1937 by Professor F. K. Kleine of the Robert Koch Institute for Infectious Diseases, Berlin. He was nominated for his work on yellow fever virus infection in mice. The professor of pathological anatomy Folke Henschen reviewed this nomination briefly. The findings were not interpreted as having sufficient originality to motivate further discussions of his eligibility for the prize.

The next nomination came in 1948, and now the development of the 17D yellow fever vaccine was the core of the proposal. The proposer was Dr Albert Sabin (page 142), who was already at that time a respected authority on the pathogenesis of viral diseases, with particular emphasis on infections in the nervous system. He had worked at the Rockefeller Institute for Medical Research between 1935 and 1939 and therefore, being at the same campus, had first-hand insight into the critical advances made in Theiler's laboratory. Sabin himself became engaged in studies of the immunological relationships between yellow fever virus and some similar viruses carried by insect vectors[24]. Thus he could well appreciate the challenge of establishing an attenuated strain of virus to be used as a vaccine. Later on he became the father of the live polio vaccine that plays a critical role in attempts to eradicate polio from our world (Chapter 5).

Sabin's nomination was very detailed, covering some six pages, and included reprints of the most important papers. The Nobel Committee of the Karolinska Institute seemed to have been impressed by the nomination and asked Sven Gard, who that same year had become professor of virus research at the Institute, to make a preliminary investigation. Gard, a pioneer in polio research, was highly qualified to make such an evaluation. He knew Theiler's work well from the ten months he had spent in his laboratory in 1939. Although Gard's investigation is referred to as preliminary, it is relatively extensive and covers eight pages. Its final two paragraphs read: "It escapes my judgment to what extent the initiative and planning of what was perhaps the most important part of the series of experiments, the initiation of the systematic tissue culture experiments, really was Theiler's. If a (further) investigation should demonstrate that the contributions by Dr (Wray) Lloyd, who died a short time after the publication of the findings, were of inferior importance in this regard, I would classify Theiler's research on yellow fever to be prize-worthy. I therefore suggest that the proposal is subjected to a separate (full) investigation." It appears that the committee was so satisfied with Gard's very thorough preliminary review that they only requested him to look further into the relative role of Lloyd in the critical experiments. In an addendum to the investigation Gard returned to this question. He came to the firm conclusion that Theiler was the leading scientist in the team and formulated this in the following way: "There was a close friendship between Lloyd and Theiler and at the time of Lloyd's death, soon after the experiments had become public, Theiler expressed his appreciation of his friend's contributions in their joint work. Lloyd is said to have had considerable energy, bordering on stubbornness, and the purposeful realization of the protracted experiments is probably in large part thanks to him. However, all the signs point to it having been Theiler who planned the experiments. His general perspective on the problems (to be solved) was already clearly formulated before they were initiated: to obtain a modified virus with reduced pathogenicity by cultivation under suitable conditions. He had also previously studied the conditions for establishment of tissue-culture-grown virus strains. The particular series of experiments discussed here is a natural corollary to Theiler's earlier work. In my opinion it is therefore probable that Theiler and not Lloyd has taken the initiative in and planned the implementation of these experiments. Information that I have received from people, who at the time worked in the laboratory, supports this opinion. On these grounds I judge it appropriate to consider Max Theiler's work on yellow fever as prize-worthy."

Gard had the proper contacts in Theiler's laboratory, dating from when he had worked there, to get advice in support of his decision on this priority issue.

The committee agreed with Gard that Theiler's contributions were prize-worthy. But the prize that year was instead awarded to Paul H. Müller for his work on DDT (Chapter 6).

Theiler's road toward the critical discovery

In his original preliminary investigation Gard first highlighted Theiler's important discovery in 1930[21,22] that yellow fever virus can be propagated by intracerebral passage in mice. Until then monkeys had been used for propagation of the virus. Repeated passages in mice led to a progressive shortening of the incubation time and, importantly, a successive reduction of the general pathogenicity for monkeys. Theiler then developed a convenient test for protecting antibodies in mice[25]. The technique allowed a quantitative demonstration of the presence of antibodies in individuals. This was an important asset for mapping the epidemiological occurrence of infections and for evaluating the effects of test vaccine products. Both a sylvan and an urban form of the disease were identified, as mentioned above. Following Theiler's work on yellow fever in mice, this animal came into widespread general use for studies of viruses, which affect humans and animals.

In the course of ongoing intensive work during the 1930s, Theiler tried to grow the virus in tissue cultures. Together with Eugen Haagen it was eventually possible to demonstrate the growth of mouse-adapted neurotropic virus in chicken embryo cultures[26,27]. The stage was now set for a full attack on the problem of establishing a stable and safe attenuated virus, which could be used without any concomitant injection of antiserum. A brute force approach was made with consideration given to the many variables mentioned above.

It was first demonstrated that the attenuation of virus obtained by passages in mice was not sufficient. The viscerotropic properties, which are the main source of the symptoms associated with yellow fever, diminished, but the neurotropic properties increased. Attempts were also made to use minimal doses of virus, but this approach also failed. Theiler and Whitman[28,29] demonstrated that, paradoxically, lower doses of virus gave a higher frequency of encephalitis in monkeys.

The comprehensive and critical experiments by Theiler and collaborators that solved the problem were performed during 1935–1937[30-32]. Different virus strains with various properties were individually carried through several hundred passages in different kinds of tissue cultures and repeatedly tested for their neurotropic activity. The breakthrough came when the Asibi strain of virus — the first one ever isolated — was passed repeatedly in minced chicken embryos from which the central nervous system had been removed. Between the 89th and 114th passage a virus variant suddenly emerged which lacked both the viscerotropic and the neurotropic effects. Fortunately the properties of this virus were stable and neurovirulence was not regained upon repeated passages of the virus in chicken embryo cultures containing brain material.

In 1936 Theiler had the so-called 17D strain injected first into himself and then into some of his collaborators. There were no complications and the development of antibodies could be demonstrated. Theiler, however, already had immunity against yellow fever from a laboratory infection at the time of the early mice experiments. He was lucky to have had a relatively mild form of the disease. Over the years there had been 32 cases of laboratory infections among employees of the Foundation and six of the collaborators afflicted had died. No cases had occurred after 1931, when early vaccine products developed by Sawyer and collaborators started to be used for immunization of laboratory personnel.

The first field trial with the new vaccine under the aegis of the Rockefeller Foundation in Brazil in 1938 was highly successful. The early experiences showed that it was important to stay as close to the original seed batch of virus as possible and also to make sure that the vaccine virus retained its infectious properties during transports in the field. The continued use of more than 400 million doses of the 17D virus vaccine for more than 60 years has shown it to be a remarkably safe and effective product. The WHO guidelines regarding the vaccine have remained unchanged for a long time[33]. Today the vaccine is still produced by means of the original methods — in embryonated chicken eggs — and stored as a frozen homogenate. At some time in the future it may become possible to produce infectious material in the form of plasmid DNA by molecular genetic techniques. This may offer a more simplified and less expensive procedure for preparation and possibly also lead to a more standardized and stable product.

One would expect that Theiler, once he had reached his goal of developing a safe and effective vaccine against yellow fever, would have experienced a eureka feeling. That may not have been the case, however. In the interview with Dr Zuckerman[17] he confirms that the achievement was satisfactory because it was successful, but the feeling remained that the yellow fever vaccine work was a deadly chore. In fact he seemed to have got a much bigger kick out of the work with mouse poliomyelitis virus[12,13], partly carried out in collaboration with Sven Gard.

During the first five years of yellow fever vaccine production at the Rockefeller Foundation human serum was the favored additive to stabilize the virus. The serum was incubated at 56°C for one hour in a water bath to destroy potentially contaminating microorganisms. Early results with some of the first batches of vaccines used in Brazil indicated that the product might be contaminated. Cases of jaundice were observed at regular time intervals after vaccination[34]. This problem was encountered on a much larger scale when it was decided to immunize all American and British troops during the Second World War with the yellow fever vaccine. The decision was taken mainly because of the perceived risk that the virus might be used in biological warfare. A total of 4.6 million doses were used in American soldiers and 1.8 million in British soldiers. The resulting number of cases of jaundice exceeded 300,000 and in some settings more than ten percent of the troops fell ill[35]. Regrettably, there was a delay in the necessary withdrawal of vaccine batches because of the hesitation of some scientists, in particular Sawyer, at the Rockefeller Foundation. They refused to acknowledge the seriousness of the problem. As the crises mounted, Theiler declared that Sawyer was "playing with fire" and he even threatened to resign from the board of the Foundation. Eventually the use of human serum was stopped and another protein stabilizer was substituted.

Many years later it was learned that the jaundice occurred because some of the sera used had been collected from blood donors carrying a hepatitis virus infection. By a supreme irony of fate, one of the most infected donors seems to have been a J. H. Morrison, the director of the Baltimore Red Cross blood donation program! The aetiology of chronic, type B, viral hepatitis was not clarified until in the late 1960s as described in Chapter 2, an achievement rewarded with a Nobel Prize in Physiology or Medicine in 1976 to Baruch S. Blumberg, a prize he shared with D. Carleton Gajdusek (Chapter 8).

An attempt to understand the vaccine-associated complications in soldiers in the wartime atmosphere of suspected biological warfare is reflected in a letter in July 1942 to John D. Rockefeller Jr. from a worried mother. A part of the letter reads: "As you perhaps know there is a great deal of jaundice in our camps, with quite a few cases proving fatal. This illness frequently follows the injection for yellow fever. I am told that the scientist in charge of this serum at your Institute is a German and many express fears that everything is not as it should be." Probably the author of the letter was thinking of Johannes Bauer, a section head of the Rockefeller Foundation at the time. However, he was a naturalized American who had been with the Foundation since the late 1920s, and who had made many important contributions, as already mentioned. By way of contrast, Theiler referred to his temporary collaborator Haagen in the early 1930s in the interview with Dr Zuckerman[17] as a German Nazi. After the Second World War Haagen was condemned as a war criminal in France for experimenting on concentration camp inmates and sentenced to 20 years hard labor. After two years he was released and returned to his laboratory at the University of Heidelberg.

Continued deliberations by the Nobel Committee

In 1949 there was no nomination of Theiler, and hence he is not mentioned in the archival documents from that year. The next time he was nominated for the prize was in 1950. Antonia Salvat Navarro from Granada, Spain, nominated him together with the above-mentioned Dr W. A. Sawyer, who for a time was the director of the laboratory where Theiler worked. The nomination is very general in its nature. It mentions the importance of developing a vaccine against the terrible disease that has afflicted many people and also scientists working with the virus. No comments are made on the special experimental advances that led to the establishment of the 17D strain of the virus. Gard, who was an adjunct member of the enlarged Nobel Committee, wrote a two-page supplementary analysis, mainly to define the relative role of Sawyer in vaccine development. Before Theiler developed his 17D strain of virus, Sawyer, together with some collaborators in the laboratory, tried to immunize humans with the previously isolated neuroadapted strain of virus from mice together with separately injected human convalescent serum. This turned out to provide some immunity, but the procedure was difficult to control and not practically useful. Gard's conclusion is that Sawyer should not be included in a prize for the yellow fever vaccine. It is said that Sawyer was shocked when he learned about Theiler's prize in 1951, since he believed he should have shared it. Less than a month after the announcement, he died.

The Nobel Committee in its final summing-up to the College of Teachers in 1950 again concluded that Theiler's work was prize-worthy. The majority of

Hilding Bergstrand (1886–1967), Vice-Chancellor of the Karolinska Institute (1942–1952). [Photo from Rolf Luft.]

the committee recommended that the prize should be given to Philip S. Hench, Edward C. Kendall and Tadeus Reichstein for their studies of hormones of the adrenal cortex. Remarkably, both Hench and Reichstein were nominated for the first time the same year (Chapter 6). However, four members, professors Nils Antoni, Hilding Bergstrand (the chairman), Gard and Arne Wallgren, out of the 13, recommended that the prize instead should be given with one half to Macfarlane Burnet for his discoveries of methods that make cells unsusceptible to certain virus infections and the other half to Max Theiler for his discovery of methods to vaccinate effectively against yellow fever. Burnet was one of the very prominent figures in the field of virology at the time (Chapter 3), but he would have to wait until 1960 before he would receive his Nobel Prize. The College of Teachers supported the majority opinion of the committee and gave the 1950 Nobel Prize to Hench, Kendall and Reichstein, whereas Theiler had to wait one more year.

On January 31, 1951 there was no proposal of Theiler for a prize. A brief nomination referring to the evaluations of the preceding year was therefore submitted that day by the chairman of the committee and the Vice-Chancellor of the Institute, Bergstrand. Such a last-minute nomination is not unprecedented in the Nobel Committee's work. If the committee notices that a particularly hot candidate, whom they want to include in the work of a particular year, has not been nominated, it can take action. Normally, however, it is the secretary of the committee who makes such nominations, which therefore are referred to as a "secretarial nomination." Such nominations are further discussed in Chapter 6.

Bergstrand had had different degrees of involvement in the work of the Nobel Committee since becoming professor of pathology in 1924. As early as 1926 he was involved in polemics with his somewhat senior colleague, professor of pathology Folke Henschen, about a possible Nobel Prize for the cancer researchers Fibiger and Yamagiwa[36]. Bergstrand was against such a prize, but regrettably he lost this fight and Fibiger received the reserved 1926 prize in 1927, one of the biggest blunders ever made by the Karolinska Institute (page 157). In spite of his involvement over the years Bergstrand had never introduced a Nobel Laureate at the prize ceremony. Time was running out, since in 1952 he was going to retire and leave all his prestigious responsibilities at the Institute. It was time to act.

Bergstrand not only made the "moonlight" nomination of Theiler, he also, somewhat surprisingly, was the one who made another evaluation. In the beginning of his four-page review he declared that he did not have anything to add to Gard's description of the events that led to the development of the yellow fever vaccine. Instead he highlighted the importance of the availability of the vaccine. He stated that it was the practical results that give Theiler an advantage in the competition with other candidates for the 1951 prize. He also expressed the hope that Theiler's success would serve as an encouragement to other scientists trying to develop vaccines against serious human virus infections.

The enlarged committee that year was composed of 11 members, but unexpectedly Sven Gard was not one of them. It is of course very meritorious to be a full member or even an adjunct member of the Nobel Committee at the Karolinska Institute and in particular to give the laudatory speech at the prize ceremony. Gard had only been a professor at the Institute for five years at the time and might have been considered junior. However, he had been an adjunct member the year before, he had written all the critical evaluations, and he was the one who knew the field of vaccinology best. So the fact that he was ostracized is telling.

In the report to the faculty all of the committee members, except two, declare that Theiler should receive the 1951 prize. The two dissenters, John Hellström, professor of surgery, and Anders Kristensson, professor of medicine, like those who had reservations about the recommendation by the committee in 1950, were harbingers of things to come. They recommended that the prize should be given to Selman A. Waksman, but this did not happen until the following year when he was rewarded "for his discovery of streptomycin, the first antibiotic effective against tuberculosis." The College of Teachers in 1951 agreed with the majority of the committee and awarded Theiler the prize. In the absence of Gard the chairman of the committee, Hilding Bergstrand, gave the *laudatio* at the prize ceremony, crowning a highly influential career at the Karolinska Institute.

Theiler received his Nobel Prize based on as few as three nominations (the early 1937 nomination did not concern the creation of the 17D strain vaccine), which may be compared with Waksman, who until 1951 had accumulated 39 nominations over six years. Out of the three nominations Theiler received, only the one in 1948 was detailed. Thus, Theiler owed a debt of gratitude to Sabin, who made this qualified proposal. These two scientists must have met many times in the late 1930s when they were both working at the campus of the Rockefeller Institute of Medical Research and had the same dining-room privileges. However, it seems less likely that they developed a closer friendship. It is hard to imagine two individuals with such diametrically different personalities. In this case it was the more humble of the two who was to be richly rewarded.

The history of viral vaccines

The development of viral vaccines represents some of the most important advances in human and veterinary medicine during the last hundred years. In fact the first viral vaccines became available before the nature of infectious agents had been identified and before any insight into the mechanisms of post-infection immunity had been gained. Edward Jenner in the late 18th century had introduced the use of cowpox virus to prevent smallpox (Chapter 2). This was the first live vaccine, which took advantage of the existence of a naturally-occurring variant of virus in cows that, through a harmless infection in man, induced an important cross-immunity. Some hundred years later Louis Pasteur developed a vaccine against rabies. He passed the agent, the nature of which he did not know, repeatedly through rabbit brain in the hope that it would change its nature. This intuitive approach was successful and the first vaccine against rabies became available. Later on this product had to be extensively modified to make it safe.

The development of viral vaccines has led to the progressive introduction of new techniques to produce either attenuated virus or safe inactivated viral antigen preparations in the laboratory. In an early phase researchers had to resort to propagation of virus in different kinds of experimental animals. During the 1920s and 1930s the first crude tissue culture techniques were introduced (see also Carrel in Chapter 6). They were difficult to work with and frequently became contaminated by other microorganisms. In spite of this Theiler managed to develop his vaccine. During the 1930s and 1940s the best source of products of certain viruses was infected embryonated hen's eggs. Influenza virus antigens produced by this approach were used for the production of an inactivated vaccine.

The major breakthrough in tissue culture techniques came in the early 1950s, when a simple technique using the enzyme trypsin to separate cells allowed the establishment of monolayer cultures (Chapter 5). Cells could be grown in large numbers and with the use of antibiotics, which now had become available (Chapters 2 and 6), contaminating microorganisms could be kept in check. Virologists could now propagate many of the medically important infectious agents. Vaccines against polio, measles and mumps, to give only a few examples, were developed. There were, however, some viruses with particular demands, which could not grow in regular tissue cultures. One example is the virus causing hepatitis B. In this special case it became possible to prepare a vaccine by use of circulating antigen recovered from blood from carriers of the infection (Chapter 2). Later on techniques were developed to produce the corresponding protective antigen by use of the recombinant DNA methodology. This technique has come into wide use to produce safe products for use in man, including several live as well as inactivated vaccines. Some vaccine products containing only DNA have also been tested.

The impact of the introduction of viral vaccines during the last hundred years is enormous. It is one of the major contributors to the success of modern health care, helping to prevent severe, sometimes crippling and deadly infectious diseases. Smallpox has been eradicated, and polio is close to being eradicated. Childhood diseases like measles, mumps, rubella and chicken pox can be prevented. They have essentially disappeared from industrialized countries, and the vaccines that can prevent them are being progressively introduced into developing countries, as means become available. Eradicating measles from our world is a realistic goal for the future. There are vaccines against respiratory infections like influenza and against enteric infections, in the form of the recently-introduced rotavirus vaccine. Finally, vaccines are also available to prevent the establishment of potentially persistent DNA virus infections. The use of the effective vaccine against hepatitis B has had major effects globally. Recently a vaccine against papilloma virus types 16 and 18 was introduced, with the capacity to prevent the development of cervical cancer in women.

Considering this long, but still not complete, list one may wonder why it is that only the yellow fever vaccine has been recognized with a Nobel Prize.

Maurice Hilleman (1919–2005), who pioneered some 40 vaccines during his long-time engagement at Merck Research Laboratories. [Private photo.]

Why were the inactivated Salk and the live Sabin polio vaccines never recognized with a Nobel Prize[37] (Chapter 5), and why was Maurice Hilleman, a giant in the field of virus vaccinology, who was personally involved in bringing more than 40 different vaccines to the market[38–40], never recognized with a Nobel Prize? It should be added that at the time of writing and publication of the shorter version of this chapter[41] the prize in Physiology or Medicine in 2008 had as yet not been awarded. This prize was given with one half to Harald zur Hausen "for his discovery of human papilloma viruses causing cancer" and the other half to Françoise Barré-Sinoussi and Luc Montagnier "for their discovery of human immunodeficiency virus." The work by zur Hausen paved the way for the development of the vaccine against papilloma viruses, but he was not himself involved in the vaccine work. So what are the criteria that the Nobel Committee and the Nobel Assembly (previously the College of Teachers) have applied — and should apply — in their deliberations?

What is a discovery?

Alfred Nobel made some interesting and important distinctions when he specified the three prizes in natural sciences in his last will. For Medicine or Physiology it states that the award should be given exclusively for "a discovery." By definition therefore all prizes in Physiology or Medicine that have been given reflect the identification of a discovery. But how can the concept "discovery" be defined and distinguished from more consequential and applied contributions? And again, why has only one of the viral vaccines been recognized, considering their enormous impact, truly in accordance with another specification in Nobel's will "to be of benefit to mankind"?

This problem is partly addressed in the next chapter discussing the award of the 1954 Nobel Prize in Physiology or Medicine to John Enders, Thomas Weller and Frederick Robbins. The breakthrough in developing a useful practical technique to grow poliovirus had important consequences for vaccine production, but the discovery had much broader implications. Its main importance was that a general technique to grow viruses of significance to man had been introduced. The major human virus pathogens were in fact isolated during the ensuing two decades. Hence we may speculate that the important contributions of Salk, Sabin and others were considered as derivative. No additional discovery was needed. So why was Theiler's contribution considered to be a discovery?

In Gard's preliminary investigation in 1948 there are some — in my opinion surprisingly harsh — comments on this matter. It reads: "Theiler ('s contributions) can not be said to have been pioneering. He has not enriched the field of virus research with any new and epoch-making methods or presented what are first and foremost new solutions to the problems, but he has shown an exceptional capacity to grasp the essentials of the observations, his own and others, and with safe intuition follow the path that led to the goal. The practical importance of Theiler's work need not be discussed. It is in large part thanks to him that the worst scourge of the tropical belt has now been rendered harmless."

In a milder form Hilding Bergstrand returns to this issue towards the end of his laudatory speech at the prize ceremony[42]: "The significance of Max Theiler's discovery must be considered to be very great from the practical point of view, as effective protection against yellow fever is one condition for the development of the tropical regions — an important problem in an overpopulated world. Dr Theiler's discovery does not imply anything fundamentally new, for the idea of inoculation against a disease by the use of a variant of the etiological agent which, though harmless, produces immunity, is more than 150 years old." One may ask if the expression "discovery does not imply anything fundamentally new" is not a contradiction in terms.

There are of course many shades to the word "discovery." In *Webster's New International Dictionary* this is illustrated by the following definitions: (a) Finding out or ascertaining something previously unknown or unrecognized; as, Curie's discovery of radium. (b) Making known; revelation; disclosure. (c) Laying open or exposing to view; unfolding. In the context of science generally, definition (a) is preferred, but there is also room for wider interpretations. In spite of Gard's and Bergstrand's comments it can be clearly argued that Theiler's scientific achievements fulfil a number of important criteria that can — and should — be applied to the term "discovery" when applied to science.

The list of qualities that may contribute to the success of a discovery can be made long and is by definition subjective. It must start with the identification of an important problem and a belief that this problem is solvable. This has been expressed well by Peter Medawar, the Nobel Laureate in Physiology or Medicine in 1960. He said: "if politics is the art of the possible, science is the art of the soluble." The problem was clearly defined, a live vaccine needed to be developed. But which approach should be taken? How useful were already available techniques and what need was there to develop new ones? In the case of Theiler he was good both at using already existing techniques and at pioneering the development of new approaches. His mouse encephalitis model, which later came to be widely applied in virology, and his modifications of the crude tissue culture techniques of the day are examples of this. His idea of growing the virus for hundreds of passages in chick embryo cultures with and without brain material was ingenious.

In the critical tissue culture experiments he showed two other personal qualities that are essential to the performance of good science. These are a capacity for a systematized approach and persistence in action. The experimental systems he used were highly complex and hence the outcome of the experiments was unpredictable. Here, two additional qualities that he apparently also displayed came to play a role together. These are his capacity to appreciate the relative significance of the many different observations made and his intuition. This is what Gard referred to in his 1948 preliminary evaluation cited above.

But still one ingredient was needed for Theiler to reach his goal. That was *luck*. He was lucky that passage of the Asibi strain in chick embryos without a central nervous system all of a sudden between the 89th and 114th serial passages changed its nature and lost both its viscerotropic and neurotropic properties, but still retained a sufficient capacity to replicate to induce an immune response. It was also fortunate that the properties of the attenuated virus turned out to be stable. Louis Pasteur's famous dictum "In the field of observation, chance only favors the prepared mind," already cited in Chapter 2, promptly comes to mind. It applies particularly well to Theiler who as a single scientist managed to make a discovery of an effective live yellow fever vaccine, with an enormous benefit to mankind.

During the last 50 years there have been immense advances in our understanding of viruses and it is now possible to completely characterize their genes and deduce the proteins they specify. In 1985 the entire nucleotide sequence of the genetic material of the 17D yellow fever vaccine virus was determined[43]. The same research group presented the complete sequence of the parental, virulent Asibi strain two years later. Differences in 32 amino acids, primarily in the surface E glycoprotein, were found. It still remains to explain the changes that are critical for the unique attenuated behavior of the 17D strain. Thus it is as yet not certain that this in-depth molecular knowledge might substitute for the empirical knowledge that Theiler so luckily gained.

Because of the unique safety and efficacy record of Theiler's yellow fever vaccine it has been studied to see if it can also be used as a carrier of critical immunogenic components of other flaviviruses and even completely distinct infectious agents[44]. By use of molecular genetic techniques chimeric vaccine candidate viruses containing surface components of the related Japanese encephalitis, dengue types 1–4, and West Nile viruses have been constructed and the early results are encouraging[45–47]. Taking the same approach, even different uniquely immunogenic structures of the circumsporozoite protein of the malaria parasite have been genetically engineered into the 17D strain of yellow fever virus[48,49]. Effective protection against parasite challenge has been obtained in experimental animals.

Finally, it would of course be of considerable interest to learn to find out what Theiler himself thought about his contribution and the fact that it was recognized by a Nobel Prize. Some insight into this is in fact possible to gain from the already-mentioned interview with Dr Zuckerman[17]. Discussing the conditions for good science, Theiler emphasized the role of hunch (possible synonyms: premonition, suspicion, intuition, inspiration, impulse) and also of luck. Being a man of paradoxes he further commented on his achievements in two contradictory ways. On the one hand, he stated that he had not done anything fundamental and that he did not have any background for making essential theoretical contributions. On the other hand he made it clear that it was he alone who had taken the decisive initiatives to the experiments that led to the development of the vaccine. Therefore in his view, if anyone should get a credit for the vaccine it should be him and him alone. No one else needed to

Max Theiler. [From *Les Prix Nobel 1951*.]

be included. Thus, although he was not a man to boast of his own achievements, he probably, in his humble way, did know his worth.

It may be appropriate to let Max Theiler himself have the last word. In his speech at the prize award banquet he used the following generous and gracious formulations[50]: "I like to feel that in honoring me you are honoring all the workers in the laboratory, field and jungle who have contributed so much, often under conditions of hardship and danger, to our understanding of this disease. I would also like to feel that you are honoring those who gave their lives in gaining knowledge which was of inestimable value. They were truly martyrs of science, who died that others might live. And, finally, I would like to feel that in honoring me you are honoring the Rockefeller Foundation under whose auspices most of the modern work on yellow fever has been done — a gesture from one great foundation to another — both having the ideal of benefiting mankind throughout the world. Thank you."

A charming encounter

When the abbreviated version of this chapter had been published[41] I was visiting Darwin Stapleton at the Rockefeller Archive Center, Sleepy Hollow, New York. He mentioned to me that he happened to know that Max Theiler's only child, his daughter Elisabeth, lived nearby. Would I be interested in meeting her? Of course, I said, and this led to a very pleasant encounter at a luncheon some time later. At this luncheon Elisabeth Martin told me about her fantastic pleasure, as a lively 12-year-old girl, accompanying her parents to the festivities in Stockholm. But she also shared with me a most charming diary that her mother Lillian had written of their visit. It repeatedly alludes to what a fantastic host Gard had been to them and it ends "Goodbye to Sven Gard who has been so kind to us."

Coda — Sven Gard's return

In connection with the 1951 prize ceremony Gard unfairly had to wait behind the scene, but his time was soon to come. He was the prime mover in ensuring the award of the 1954 prize to John Enders and collaborators for the growth of polio virus in non-nervous cultures[37]. He did all the critical evaluations, convinced the faculty not to follow the majority of the committee and gave the laudation at the prize ceremony. His close familiarity with the field of polio research allowed him to make a fair judgement on the most influential actors in developments that led to the production of an effective polio vaccine. This vaccine allowed the elimination of a disease that had developed as a major threat to health in industrialized countries. This success story, highlighting in a perfect combination the critical requirements of Nobel's will — "discovery" and "benefit to mankind" — will be the theme of the next chapter.

Chapter 5

Polio and Nobel Prizes

In 1954, John Enders, Thomas Weller and Frederick Robbins at the Harvard Medical School, Boston, were awarded the Nobel Prize in Physiology or Medicine "for their discovery of the ability of poliomyelitis viruses to grow in cultures of various types of tissue." This discovery provided for the first time effective means to produce both inactivated and live polio vaccines. It was also of major importance for facilitating the growth of many different medically important viruses.

Polioviruses are small enteric RNA viruses that on rare occasions spread to the central nervous system (CNS) and cause disease. Replication of the virus in anterior horn cells of the spinal cord can result in paralysis. In the 1950s, both inactivated and live polio vaccines were developed to prevent paralytic disease. Three vaccines for types 1, 2 and 3 polioviruses had to be developed to create effective immunity.

Although polio has been largely eradicated using primarily live vaccines, it persists in some parts of the world. The goal of global polio eradication, like that achieved for smallpox in 1978, remains elusive. Target dates have been moved back repeatedly. Still, most public health officials retain their belief that it is possible to eradicate polio from our planet using a combination of inactivated and live poliovirus vaccines. In fact, the type 2 strain of the virus has been eradicated. The World Health Organization assisted by a number of non-governmental organizations is now aiming at eliminating the last reservoirs of poliovirus types 1 and 3. Presently, wild viruses of these types remain endemic in four countries. From these areas, they can spread to reinfect people living in countries that were previously declared to be polio-free.

The global campaign of immunization against poliomyelitis was made possible by two vaccines: an inactivated virus preparation developed by Jonas Salk, Julius Youngner and their colleagues[1-3] and live, attenuated virus preparations developed initially by Hilary Koprowski, Herald Cox and their co-workers[4-6] and later by Albert Sabin and his colleagues[7]. These vaccines dramatically changed the lives of millions of children in developed countries where polio vaccination was initially introduced. Examples of the extraordinary effectiveness of the vaccines are illustrated by the decline in poliomyelitis cases in the US and Sweden (see figure on page 126).

By any measure, the eradication of polio must be considered a milestone in the annals of medicine. That said, it is reasonable to ask why the successful development of both inactivated and live polio vaccines was not celebrated by a Nobel Prize. First, it can be argued that Alfred Nobel would have thought such work, which improved the lives of so many, would be most appropriate for the award that bears his name. Second, one may ask why the Nobel Prize was awarded to Enders, Weller and Robbins in the fall of 1954 just when the first clinical trial of the Salk vaccine was being completed. Why didn't the Nobel Committee wait to view the outcome of a mass immunization program encompassing nearly 650,000 children that was announced in the spring of 1955[8]? In the hope of answering the questions posed above I and Stanley Prusiner (Chapter 8) used the Nobel archives at the Karolinska Institute to investigate the circumstances surrounding the award of the 1954 Nobel Prize[9]. Before presenting the results of this examination some of the salient features of polio infections are reviewed.

Polio epidemics in the 20th century

The poliovirus seems to have infrequently caused CNS disease in humans prior to the 19th century although atrophied limbs in young people were recorded in ancient Egypt[10]. Presumably, non-symptomatic enteric infections of the poliovirus in young children were so common that these conferred widespread immunity. These young individuals suffered no CNS dysfunction because they were protected by maternal antibodies remaining from the time of delivery. As hygiene improved and public sanitation measures

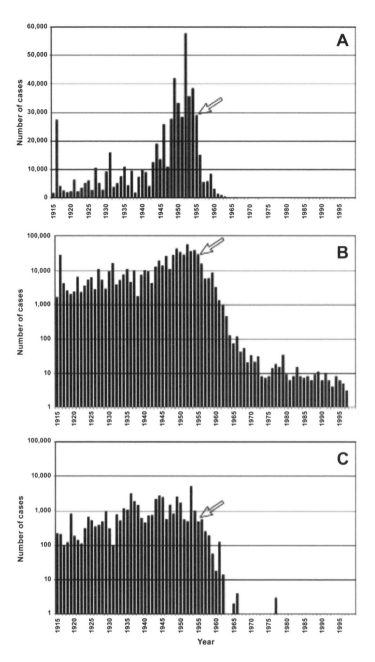

Epidemiology of annual number of polio infections in the US (A, B) and Sweden (C) between 1915 and 1999. Polio disappeared from the US and Sweden after the introduction of inactivated vaccines (arrows). Data are shown in arithmetic (A) and logarithmic (B, C) scale. Note that Sweden eradicated polio by use of only inactivated vaccine whereas the US, like most other countries, switched to the live vaccine in 1963. The incidence of polio cases in the US until 1999, when the inactivated vaccine began to be used again, is due to disease associated with the live vaccine and to imported cases. Data were provided by Margareta Böttiger, Michael Katz and Post-Polio Health International.

were implemented, the age at which children developed their own antibodies increased. Since maternal antibodies to poliovirus in children disappear between the ages of 1 and 2 years, an increasing population of susceptible children emerged. These young people had not acquired immunity to polioviruses through unapparent enteric infections as infants when they were still protected by their mother's anti-polio antibodies[11]. Gradually a non-immune population of older children and young adults started to become represented and when exposed to polioviruses, many of them developed paralytic polio.

The first well-described outbreaks of polio were registered among children in Scandinavia. Heine, a German orthopedist, gave the first complete clinical description of the disease in 1840 and Medin, a Swedish pediatrician, reported the first epidemics[10]. The number of outbreaks grew with time and they began to spread to other geographical areas. The disease originally known as Heine-Medin's disease and later, infantile paralysis or poliomyelitis, is today generally referred to as polio. The latter name of the disease derives from the prominent damage of the nerve cells in the gray matter of the spinal cord; Greek *polio* — gray — and *myelitis* — matter/marrow (*itis* is a Latin suffix meaning inflammation).

During the first half of the 20th century, the incidence of polio increased and epidemics in the summer months became commonplace in many industrialized countries (see page 126). Year after year, the epidemics killed children and left many more crippled. The disease created anxiety, horror and political unrest. In the US, efforts to deal with the recurrent polio epidemics were spearheaded by Franklin Roosevelt, who himself was crippled by polio in 1921, and by Roosevelt's associate Basil O'Connor. When Roosevelt was elected President of the US in 1932, he was in a position to acquire and dedicate enormous resources to a national campaign against polio. Roosevelt and O'Connor created the National Foundation for Infantile Paralysis and raised immense funds through annual campaigns called the "March of Dimes" that were used to care for polio victims and support vaccine research. Today, the National Foundation supports research on birth defects.

Initial attempts to produce a vaccine

In 1908, Karl Landsteiner demonstrated the viral origin of poliomyelitis when he transmitted the disease to monkeys by using a filtered preparation of

Karl Landsteiner (1868–1943), recipient of the Nobel Prize in Physiology or Medicine 1930. [From *Les Prix Nobel 1930*.]

macerated CNS tissue obtained from individuals who had died of polio[12]. Landsteiner later discovered human blood groups, for which he was awarded the Nobel Prize in 1930. The filterable agent causing poliomyelitis was difficult to study since there were no simple procedures for growing the virus in the laboratory. Experiments had to be performed with tissues harvested from infected monkeys.

By using inactivated virus from the brains of polio-infected animals, it was possible to demonstrate that effective immunity could be induced in monkeys[10]. It was also possible to demonstrate, by rather cumbersome but important experiments with monkeys, that three different types of polioviruses are distinctly antigenic. This meant that not one, but three different vaccines had to be developed for polioviruses types 1, 2 and 3. Additionally, the idea of preparing enough vaccine for massive immunizations was considered not feasible for two reasons: first, the number of monkeys needed would be enormously expensive and secondly, the presence of non-poliovirus antigens might cause damage to brain tissue.

In the 1930s and 1940s, several attempts to develop a polio vaccine resulted in abysmal failures. The failures sensitized the medical community to the disastrous results that an ineffective vaccine could bring. By the mid 1930s, transmission of the Lansing type 2 strain of poliovirus from humans to mice was reported[13]. A few years later, transmission of the Lansing strain from monkeys to cotton rats was described[14,15]. But these results were not pursued until the late 1940s when Hilary Koprowski and co-workers, using the poliovirus attenuated in the brains of cotton rats, produced an experimental live virus vaccine and thus initiated the modern era of polio vaccinology. In 1950, Koprowski orally vaccinated himself, his technician and, later, a group of children with an extract prepared from the brains of infected animals[4]. A critical breakthrough in vaccine development occurred when it was established that polioviruses could be grown in non-nervous tissue[16-19]. This discovery resulted in the award of the 1954 Nobel Prize in Physiology or Medicine to Enders, Weller and Robbins. In 1956, Koprowski switched from vaccines prepared from attenuated poliovirus strains propagated in rodent brain to those grown in cultured monkey kidney cells[20,21].

Early deliberations by the Nobel Committee

Nominees for a Nobel Prize in Physiology or Medicine are evaluated at three levels: (1) short notes for relatively weak candidates, (2) preliminary reviews of a few pages for stronger candidates, and (3) exhaustive analyses for the strongest candidates. Generally, a preliminary review precedes a detailed analysis. In addition to documents of this kind there is a record of the concluding meeting of the enlarged Nobel Committee (at the time the nominal three-member committee with adjunct members chosen annually). This record is a decision document; it presents a list of the major candidates, comments on their prize-worthiness and gives the proposal for the prize recipients of the year reviewed. Nowhere in the archives, except for some of the first years of the Nobel work at the Karolinska Institute, can one find the opinions of the individual committee members with respect to particular nominees as mentioned before.

Enders was first nominated for a Nobel Prize by Dr L. P. Gebhardt in 1952, for his discovery that polioviruses could be propagated in cultures of non-nervous tissues. Only Enders was named. He was considered such a strong candidate that an exhaustive review was made. In a document of over 20 typewritten pages, Sven Gard described the background for this discovery and the dramatic changes in poliovirus research that it had brought in the past five years. The practical consequences of Ender's discovery were also highlighted in this review, much of which is summarized below. Before the discovery by the Enders' group, many attempts had been made to grow polioviruses in tissue cultures. 40 years earlier, Simon Flexner who was the powerful first director of the Rockefeller Institute, New York, introduced in the previous chapter, had reported his attempts to achieve this. Based on observations in infected monkeys, Flexner and his colleagues concluded that poliovirus replication occurred exclusively in neural tissue or cells[22,23]. Two decades later, Albert Sabin and Peter Olitsky reinvestigated the growth of poliovirus in tissue culture using the neurotropic monkey-adapted MV strain provided by Flexner[24]. They interpreted their data as confirming Flexner's earlier findings.

Gard commented on the misconception that polioviruses could not be grown in non-nervous tissue cultures when he wrote:

Sabin's and Olitsky's work was viewed for some 15 years to come as the last word on in vitro replication of poliomyelitis virus. It was concluded on the basis of their data that virus, not only in vivo, but also in vitro displays a pronounced species specificity and an extreme neurotropism.

The widely held perception that poliovirus replication was confined to nervous system tissue was overthrown by the pioneering work of Enders, Weller and Robbins[16-19].

Enders, viruses and cultured cells

Enders had a longstanding interest in growing viruses in tissue cultures. In 1940, he engaged a medical student, Weller, for a tutorial research project. Together with another researcher, Dr Alto Feller, they managed to obtain substantial replication of vaccinia virus in chicken embryo tissue maintained in a roller tube culture system[25]. Weller left to serve in the armed forces in World War II and did not return to Enders' laboratory until 1946. About a year after Weller's return, Robbins, a medical school classmate, joined him in Enders' laboratory. Incidentally, Robbins was the nephew of John H. Northrop, the Nobel Laureate who was presented in Chapter 3. Enders asked Robbins to review the available tissue culture techniques for their use in the propagation of viruses and assigned him the task of growing viruses from children with infant diarrhea ("The Reminiscences of Robbins" (April 2, 1964), in the Oral History Collection of Columbia University). Meanwhile, Enders tried to cultivate the measles virus in tissue culture and Weller, the chickenpox virus. In addition to measles and chickenpox viruses, Enders and his colleagues examined the ability of other viruses to grow in tissue cultures. In 1949, Weller and Enders published a manuscript on the successful cultivation of mumps and influenza viruses in suspended cell cultures with the production of hemagglutinin[26]. By the early 1950s, they successfully cultivated half a dozen human viruses in a series of pioneering studies[27,28].

In spite of the fact that the main focus of the laboratory was not to grow polioviruses, a series of experiments, performed between March and June 1948, demonstrated the robust growth of such viruses in tissue culture. In his attempts to isolate varicella virus, Weller used tissues obtained from aborted human embryos. Some unused cultures that had been established for another set of experiments were inoculated with the cotton rat-adapted Lansing strain of poliovirus. Although no growth of varicella virus was recorded in some cultures, poliovirus grew spectacularly. Using bioassays in rats, extremely high amounts of poliovirus were documented in these cultures. The reasons for the inoculation of poliovirus into the leftover tissue cultures, and who took the initiative, are not clear. In a retrospective, Weller[29] stated that he initiated the poliovirus experiments, but according to Robbins[30,31], the idea for these studies came from Enders.

Sven Gard and his role

Discussions about awarding a Nobel Prize for the growth of polioviruses in cultured cells evolved over three years, from 1952 to 1954. As a prominent researcher in the field of polio and polio-related viruses, Sven Gard was unusually well qualified to lead such discussions. Whether he was so opinionated that he should have excused himself from such important deliberations is unclear — certainly, his encyclopedic knowledge of polioviruses must have been considered by many to be a great asset. Gard was a physician who was fascinated by microbial diseases. He appeared to have developed an interest in polio through contacts with Carl Kling. That polio might be a waterborne disease, which is spread by sewage-contaminated drinking water, had already been proposed in the early 1900s by Kling[10]. At the time this was considered a very controversial proposal.

In 1939, Gard worked at the Rockefeller Institute as a visiting scientist in Max Theiler's laboratory. The Nobel Committee was familiar with the importance of viral vaccines. As described in the previous chapter Theiler received the 1951 Nobel Prize in Physiology or Medicine. Gard did not become involved in Theiler's yellow fever studies, but collaborated instead on investigations of a murine, polio-like virus known as the Theiler virus[32,33]. Gard also participated in epidemiological studies of human polioviruses in the US. Just before the Second World War erupted Gard managed to return to Sweden. Back home he pursued his interest in polio and polio-like viruses under the guidance of professors The Svedberg and Arne Tiselius at Uppsala University, which were introduced in Chapter 3. By 1943, Gard completed his PhD thesis on the purification of the murine polio-like virus from the brains of tens of thousands of mice by the use of physical-chemical methods available at the time[34] as mentioned earlier.

In the mid 1950s, Gard and his collaborator Erik Lycke pioneered an understanding of the kinetics of formalin inactivation of poliovirus. Salk had argued that the inactivation was linear[3,35], that is, that the log of remaining virus

Swedish polio pioneers Sven Gard (left) and Erik Lycke (right).
[Photos, private and from Erik Lycke.]

activity was a linear function of the time of treatment with formaldehyde, but Gard and Lycke showed that inactivation did not proceed according to a simple first-order reaction, indicating that the interactions between the virus and formaldehyde were more complicated[36-39]. This finding indicated that it was not possible to estimate the length of the formaldehyde treatment merely by extrapolation from a linear inactivation curve. These considerations appear to have been quite pertinent in explaining the tragic "Cutter incident" in the US in 1955, in which more than 200 cases of polio occurred in children receiving one of the early batches of the inactivated vaccine as well as in their families and community contacts[40-44]. At a symposium in London in 1957, Gard presented an empirical formula fitting the available experimental data[45]. Gard and Lycke published a thorough analysis of calculations using Gard's formula and applied these to the results of both Swedish and German studies. The work of Gard, Lycke and their colleagues led to the introduction of modified conditions for the production of an inactivated polio vaccine in Sweden[46,47]. The positive experiences of the modified, inactivated vaccine resulted in Sweden never resorting to use of the live vaccine. The eradication of polio in Sweden was achieved solely with inactivated preparations[48,49].

Nomination of Enders in 1952

As mentioned above, one Nobel Prize nomination was submitted in 1952 for Enders alone. In his analysis of Enders' work for the Nobel Committee in 1952, Gard pointed out that the dogma of strict neurotropism of the poliovirus was beginning to be questioned by the late 1940s. After high levels of poliovirus were found in human feces, investigators began to question how so many virions could be produced by nerve endings in the intestinal mucosa[50]. A more likely explanation was that polioviruses replicated in nonneural tissues, an interpretation that Enders seemed to embrace. Moreover, Enders encouraged Robbins to pursue additional experiments on the growth of poliovirus in tissue culture once the initial results were obtained[16-19]. The last paragraph of their *Science* paper[16] references the dogma of poliovirus neurotropism in light of their new findings. It reads:

> It would seem, from the experiments described above, that the multiplication of the Lansing strain of poliomyelitis virus in the tissues derived from arm or leg, since these do not contain intact neurons, has occurred either in peripheral nerve processes or in cells not of nervous origin.

Gard emphasized in his evaluation that Enders and collaborators did not invent any new tissue culture technique. In spite of this, they were successful in propagating poliovirus where other investigators had failed. At least two explanations might account for this difference in the results. First, choosing the Lansing poliovirus strain may have been critical. Sabin and Olitsky had used the MV strain of poliovirus in their attempts to propagate the virus in tissue cultures[24]. The MV strain is a highly neurotropic virus that was established by Flexner during 20 consecutive passages in monkey brain and is likely to have a reduced capacity for growth in non-nervous cells[51,52]. The Lansing strain used by Enders, Weller and Robbins was also neurotropic, but passage in this case had occurred in cotton rats[14,15]. Whether repeated passage through monkeys versus rats is the correct explanation for this difference in growth properties is unclear. Second, the Enders group allowed a longer time for the virus to grow in cultures. Sabin and Olitsky split their cultures every third day according to the conventions of the time, whereas the Enders group kept the cultures for weeks with repeated changes of the culture media in order to renew the nutrients.

In 1952, Gard came to the conclusion that the growth of poliovirus in non-nervous tissue was worthy of a Nobel Prize, but he "refrains at present from formulating an opinion on whether a Nobel Prize in this field should be given to Enders alone." The Nobel Committee agreed that

Enders' contribution was worthy of a Nobel Prize, but the prize that year was awarded to Selman Waksman "for his discovery of streptomycin, the first antibiotic effective against tuberculosis." It is notable that antibiotics, such as penicillin and streptomycin added to tissue culture media greatly facilitated the recovery of replicating viruses in the cells by suppressing bacterial contamination.

Nominations of Enders and collaborators in 1953 and 1954

In 1953, Enders was again nominated for the Nobel Prize: this time by Dr J. H. Means and by Dr John H. Dingle. No further analysis of Enders was made that year and the Nobel Committee did not mention him in their summary report. Curiously, Gard was not an adjunct member of the Committee in 1953 but he is said to have visited Enders' laboratory in Boston in October 1953 (T. Weller, personal communication, 2005). Whether the impressions that he gained during this visit influenced his interpretation of the relative contributions of Enders, Weller and Robbins, remains unknown. This contact between Gard and the Enders laboratory at Children's Hospital in Boston is potentially the source of the widely circulated, but unsubstantiated, story that Enders contacted the Nobel Committee stating that he would refuse to accept the Nobel Prize unless his two young collaborators were included[53]. This story has been widely circulated and Prusiner, the collaborator in this review of the archives, remembers learning about Enders' attempt to influence the Nobel Committee in a medical school lecture at the University of Pennsylvania. It has not been possible to find any correspondence between Enders and the Nobel Committee, much less a letter stating the conditions under which he would be willing to accept the Nobel Prize. In this context, it should be added that Enders accepted both the Passano and Lasker awards alone prior to the award of the Nobel Prize.

In 1954, nine nominations of Enders, several of them relatively exhaustive, were submitted for a Nobel Prize. Some of these came from prominent scientists in the field of virology: John R. Paul, Christopher H. Andrewes and F. Macfarlane Burnet. Burnet had already been nominated yearly since the late 1940s for his contributions to the field of virology (Chapter 3) and in 1960 he shared the Nobel Prize with Peter Medawar for an important discovery in the field of immunology. Only two of the nine nominations of Enders included Weller and Robbins. These two nominations came from less authoritative researchers in the field: P. L. Lence from Ljubljana, Yugoslavia and Dr G. Bruynoghe from Louvain, Belgium. Bruynoghe's proposal also listed other nominees, including G. K. Hirst, D. Horstmann and D. Bodian. Lence, in his nomination, referred to "the production of a non-toxic polio vaccine." In his letter, he first cited the successful growth of polioviruses in non-nervous tissue by "Enders *et al.*" Lence stated, "By this discovery, the main hurdle to production of a vaccine was passed." He then noted as follow-up information on the cultivation by Cox[54,55] of poliovirus in chicken embryo cells and the first attempts to produce a vaccine by Koprowski *et al.*[4] and the work of "Youngster [Youngner] *et al.*" Lence concluded by recommending that Enders, Weller, and Robbins should receive the Nobel Prize. The names of Bruynoghe and Lence are not known to Erik Lycke (personal communication), the close collaborator of Gard in the early 1950s, and hence probably not to Gard himself at the time. They may have been microbiologists, but they were unlikely to have been virologists. Since their nominations were accepted, they must have been affiliated with an academic institution that was invited to make proposals by the Nobel Committee for Physiology or Medicine[56]. What would have happened in 1954 if their nominations had not been submitted remains uncertain.

The Nobel archives reveal that in 1954, Sven Gard wrote another review of the developments in the field but this one was only five typewritten pages in length. In the beginning of this review, he stated: "It seemed likely to me that both Weller and Robbins had taken active part in the planning and execution of the experiments." He emphasized the rapid development of the field, with possibilities for providing laboratory support to clinical diagnoses and epidemiologic surveillance, as well as for production of large quantities of virus for vaccine purposes. He clearly noted that:

> In preliminary experiments with many thousand individuals, Salk has demonstrated that formalin-inactivated virus can produce a considerable serological immunity. At present, field trials are performed on a large scale to assess the protective efficacy of the Salk vaccine. It was tested in a total of 650,000 children in the USA, 25,000 in Canada and 20,000 in Finland, out of which about 1/3 have received an inactivated control preparation. The results of these trials are not expected to become available until the beginning of next year.

Gard concluded his evaluation with an extraordinarily enthusiastic assessment:

It is not an exaggeration to state that the discovery by Enders' group is the most important in the whole history of virology … . The discovery has had a revolutionary effect on the discipline of virology.

In the last paragraph of his analysis, Gard reiterated the prize-worthiness of the discovery and concluded,

Since the time when I submitted my previous evaluation I have come to the firm conviction that no one of the three members of the group can be said to have contributed more than any of the others to provide a solution to the problem. If the decision is taken to reward the discovery with a Nobel Prize, which I consider to be highly appropriate, I would propose that the prize should be given jointly to Enders, Weller and Robbins.

Interestingly, the names are given in the above order, and included thus in the Nobel Foundation Directory, and not in alphabetical order as listed below.

The decision

The conclusion of the enlarged Nobel Committee in the document sent to the College of Teachers of September 28, 1954 reads:

The Nobel Committee decided to propose that the 1954 Nobel Prize in Physiology or Medicine should be given to Vincent du Vigneaud for his discovery of the structure of vasopressin and oxytocin, confirmed by the synthesis of these hormones. Professors Gard and Hellström were of the opinion that the prize instead should be given to John Franklin Enders, Frederick C. Robbins and Thomas H. Weller jointly for their discovery of the capacity of poliomyelitis virus to grow in different tissue cultures from primates.

Expression of dissenting views in the final proposal from the committee is a relatively uncommon phenomenon. Generally, the committee attempts to make a unanimous recommendation. During the ensuing debate by the College of Teachers, apparently Gard and Hellström managed to swing the opinion of the majority in favor of Enders, Weller and Robbins, so that they became recipients of the 1954 prize. This was neither an isolated nor precedent-setting situation; other historical examples exist in which the College of Teachers of the Karolinska Institute did not follow the recommendation of its Nobel Committee.

Secrecy surrounding the selection of Nobel Prize recipients is generally well-maintained but in 1954, the recommendation of the Nobel Committee of the Karolinska Institute was leaked to a US newspaper[57,58]. *The New York Times*, but not the main Stockholm newspaper of the day, reported "Up to the time of voting, the faculty (of the Karolinska Institute) was more or less divided between the selected trio (Enders, Weller, and Robbins) and another American, Professor Vincent du Vigneaud, age 53, of Cornell and New York." du Vigneaud must have been very disappointed when he learned of his failed candidacy. However, he was compensated the following year when he received the Nobel Prize in Chemistry.

Thomas Weller (left), Frederick Robbins (center) and John Enders (right). [Photo from Reference 29.]

In the fall of 1955, both the Chemistry Nobel Committee of the Royal Swedish Academy of Sciences and the Nobel Committee at the Karolinska Institute recommended Hugo Theorell for a Nobel Prize. Since the faculty of the Institute had its meeting first it secured that Theorell, one of their own, received the prize in Physiology or Medicine. The members of the Royal Swedish Academy of Sciences at its later meeting therefore had to take number two on their list for the prize in Chemistry. That was du Vigneaud. Incidentally, Theorell was a victim of polio and used a cane to walk.

Polio research and vaccine production

Before Enders and his colleagues left the field of poliovirus propagation and vaccine production, they demonstrated that the neurovirulence of poliovirus could be attenuated by repeated passage in tissue culture[59]. Enders later used the same approach with the measles virus to generate the strain that in a further attenuated form remains in use today as a vaccine[60].

The ability to grow poliovirus to high titers in tissue cultures set the stage for development of effective, safe vaccines. Following the lead of Enders and co-workers, Salk and his collaborators showed that monkey kidneys provided a useful substrate for large-scale production of poliovirus and for preparation of a formalin-inactivated vaccine[1,2,61]. In this work, Julius Youngner, who was a member of the Salk group, introduced an important technical improvement when he, like Dulbecco and Vogt[62], resurrected the method of trypsinizing tissue fragments initially employed at the Rockefeller Institute[63]. Monolayer cell cultures were established using the trypsin technique and became the standard for most future work[64,65]. By recording changes of the appearance of cells — the cytopathic effects, a term introduced by Enders and his colleagues — numerous medically important viruses were identified during the 1950s and early 1960s.

The growth of poliovirus in tissue culture facilitated both the isolation of attenuated strains that were suitable for live vaccines and the large-scale production of these attenuated viruses that enabled mass vaccination programs. Studies with tissue culture-grown, attenuated poliovirus begun around 1953, but the vaccines were not recommended for general use until 1961. Parallel studies of competing live vaccines were undertaken by Koprowski[20,21], Cox[5] and Sabin[7,66]. Eventually, the three vaccine strains developed by Sabin became the live vaccine of choice because they were thought to give the lowest frequency of vaccine-associated cases of polio.

Under the aegis of the National Foundation for Infantile Paralysis in the US, the inactivated vaccine developed by Salk came into general use after a successful initial trial in 1954, which led to a dramatic reduction in the occurrence of polio cases (page 126)[8]. In spite of this success, the Salk vaccine was replaced by the Sabin live vaccine in 1961; the latter vaccine was easier to administer and presumably had an improved capacity to induce herd immunity because of the spread of attenuated virus from vaccinated individuals. Eventually, after decades of use, the Sabin live vaccine was replaced in 1999 by the Salk inactivated vaccine that had been used originally in the US. Because reversion of the attenuated polioviruses to wild-type occurs in vaccinees at a low rate, use of the Sabin vaccine is no longer recommended in countries where polio has been almost eradicated.

Why not wait for the first vaccine trials?

While the discovery of Enders, Weller and Robbins was critical to polio research, virology and future vaccine development, it is reasonable to question why the College of Teachers of the Karolinska Institute awarded the Nobel

Prize in Physiology or Medicine to Enders, Weller and Robbins in the fall of 1954. Why did they not want to wait to learn the results of one large and two smaller polio vaccine trials that had been started in the spring of that year? After all, the critical deliberations took place about four months after all the children in these vaccine trials had been immunized.

Not surprisingly, Jonas Salk was nominated for the first time in 1955 for the Nobel Prize in Physiology or Medicine. At the time of nomination, the results of the large field trial, mentioned in Gard's 1954 evaluation of Enders and collaborators, were still pending[8]. The rigorous analysis offered by Thomas Francis on April 12, 1955, showed the Salk vaccine was clearly protective. The incidence of polio among the almost 200,000 children who had received the vaccine was reduced at least 50 percent with no adverse side effects reported after vaccination. In response to Salk's nomination by Drs A. J. Carlson and H. A. Rusk, Sven Gard wrote what appears to be a rather ambiguous formulation in a preliminary evaluation submitted on April 13, 1955:

> It appears to me that the problem (of polio vaccine production) is of such importance from a practical medical viewpoint that a more comprehensive review is appropriate. However, it is hardly possible to take a conclusive position for the moment. The results of the field trials that last year were conducted in the USA, Canada and Finland to a major extent have an influence on the standpoint to be taken. The results of these trials have now been compiled, but the complete report will not be available for some time to come. Under these conditions I *still* consider that I should propose that the work is subjected to an exhaustive analysis.

However, the committee did not initiate any in-depth review. In response to three nominations of Salk the next year by Leslie A. Osborn, Karl T. Neubuerger, and A. Sarpyener, Gard prepared another preliminary analysis consisting of eight typewritten pages. Gard described the first animal immunizations in 1910, studies of six different procedures for inactivation of polioviruses, the failed immunizations in the 1930s by Kolmer and Brodie as well as Flexner's view that inactivated poliovirus is not immunogenic. Next, Gard comments on studies showing three distinct poliovirus strains, each of which requires a separate vaccine.

With this background, Gard analyzed Salk's contributions. He described Salk's faulty interpretation of inactivation studies that were used to define conditions for the manufacture of polio vaccines. Gard argued that Salk's rigid attitude and incorrect recommendations were responsible for the Cutter incident. He concluded:

> Salk's most important contribution is according to my opinion that he definitely demonstrated that serological immunity and protective effects can be obtained by use of a formalin inactivated poliovirus vaccine. This is in principle nothing new and furthermore Salk has not in the development of his methods introduced anything that is principally new, but only exploited discoveries made by others. It has not been possible to reproduce some of his experimental results in other laboratories and it seems now reasonably well secured that some of his working hypotheses are in fact incorrect. It cannot be excluded that some of the accidents that occurred in connection with the mass immunizations in the US in 1955 result directly from the practical application of such incorrect hypotheses. According to my opinion Salk has not demonstrated the cautiousness that one would expect to be applied in this context. It is my view, based on these conclusions, that Salk's publications on the poliomyelitis vaccine cannot be considered as prize-worthy.

In 1958 Koprowski and Sabin were jointly nominated for their work on live polio vaccines. This was the first time they had been proposed for a Nobel Prize in Physiology or Medicine. Gard made another insightful preliminary four page evaluation. He noted the different techniques used by the two scientists to develop attenuated virus and he preferred the more systematic genetic approach used by Sabin. Obviously Gard was not in principle negative to the use of live vaccines, since he himself was involved in the testing of both Koprowski's and Sabin's candidate strains in Sweden. The conclusion of the report summarized that the contributions "... represent very important steps in the fight against polio. However, it is not possible at the present time to make a firm statement on their practical value." Possible that more firmly supportive future evaluations of the developments of live polio vaccines remained hidden in the penumbra of unavailable Nobel archives at this time.

In the late 1960s, an initiative emerged from Rune Grubb, professor of bacteriology at Lund University in Sweden, to nominate Salk, Sabin and Koprowski for the Nobel Prize (E. Lycke, personal communication). Among the long list of names, numerous Swedish microbiologists signed the petition.

In this initiative, Gard was included as a fourth nominee. The nomination cited Salk and Gard for the development of the inactivated vaccine as well as Sabin and Koprowski for the attenuation of poliovirus strains and development of live polio vaccines. After wide discussions, the nomination was signed by professors of virology, bacteriology and immunology from many universities in Sweden. It is likely that professors from other Scandinavian countries also signed the nomination. When the nomination came to Gard's attention, without a moment's hesitation, he made it clear that he would not accept any nomination. He justified his refusal by referring to the Nobel statutes that the prize was to be given for achievements of a primary nature and not for applications of work derived from the accomplishments of those already awarded the prize. Certainly, the long road to eradicating polio through the use of either inactivated or live vaccines undoubtedly include remarkable advances of science, some of which might have qualities of a "primary nature." But, Gard's decision was final and the nomination was never submitted to the Nobel Committee. The nomination of four scientists, when the rules of the selection of recipients of Nobel prizes clearly stated a maximum of three awardees, was flawed from its inception.

Returning to the question of why late in the fall of 1954 the College of Teachers chose to support a minority opinion of the Nobel Committee and awarded the Nobel Prize in Physiology or Medicine to Enders, Weller and Robbins: one can only speculate about their motive. As the most knowledgeable person on the subject of polioviruses in Sweden at this time, Gard was in a position to speak authoritatively. His evaluation for the Nobel Committee shows the admiration that he had for the discovery of the Enders group when he described their work as "… the most important in the whole history of virology." Such hyperbole might best be understood within the context of Gard's presentation speech at the Nobel Prize award ceremony on December 10, 1954[67]. In the final section of his comments addressed to Enders, Weller and Robbins, unquestionably, Gard understood the importance of their discovery for future vaccine production, but for him, their work had much wider implications. Clearly, Enders, Weller and Robbins laid the foundation for the accomplishments of many virologists in the 1950s and early 1960s. During that period, the majority of medically important human viruses were identified by examination of cytopathic effects in tissue cultures.

It can be concluded that Gard's admiration for the work of the Enders' group prompted him to push for the award of the 1954 Nobel Prize and that his persuasive personality greatly influenced the College of Teachers. Gard's

Polio vaccine pioneers: (A) Herald Cox, (B) Hilary Koprowski, (C) Albert Sabin, (D) Jonas Salk, and (E) Julius Youngner. [Courtesy of Indiana State University, Samuel Katz, Hilary Koprowski, and Julius Youngner.]

caustic analysis of Jonas Salk's nomination in 1956 suggests that any future nominations of him, and possibly also other polio vaccine pioneers, were doomed. It can be mentioned that Salk received ten nominations in 1958 and seven nominations the following year. Certainly, it is reasonable to assume that until 1972, when Gard retired, no nomination of Salk was ever considered seriously by the Nobel Committee.

Because of the impact of his contributions Salk became one of the scientists best known to the general public. He developed into a self-appointed philosopher and had a dream of establishing an environment where humanistic and natural sciences would meet. Considerable resources were raised and together with the architect Louis Kahn a beautiful building, which today has become a landmark, was raised in La Jolla overlooking the Pacific Ocean. The project was finished in

1965. Salk's dream may not have been fully realized but the Institute carrying his name has evolved to become a leading research center for genetics, molecular biology and neurosciences (Chapter 7). In 1969 Salk met Françoise Gilot, a French painter and best-selling author of the book *Life with Picasso*. They married in 1970. Throughout the last 15 years of his life I had frequent contacts with Salk. Over many meals he liked to reflect on philosophical issues and the directions of his thoughts can be seen from his books with titles like *Man Unfolding, The Survival of the Wisest* and *Anatomy of Reality: Merging of Intuition and Reason*. Salk knew his value and concluded from his overwhelming recognition by society that he was an anointed individual. Still there was always an undertone of disappointment in our discussions because of the fact that he was not fully recognized by his scientific colleagues. He never became a member of the US National Academy of Sciences.

Coda — A unique contact

There is a regulation that scholarly investigations of archive materials of prizes in Physics and in Chemistry are not allowed if the prize recipients are still alive. This rule does not apply to materials concerning the prize in Physiology or Medicine. Hence, Prusiner and I could investigate the 1954 prize although one of the awardees, Thomas H. Weller, was still alive. He died in 2008. I corresponded with him in 2005 and he kindly sent me a letter with a dedicated copy of his autobiography *Growing Pathogens in Tissue Cultures. Fifty Years in Academic Tropical Medicine, Pediatrics and Virology*[29] and some additional material enclosed. He asked me not to cite his letter, since he, as the last survivor of the three laureates, did not want to expose his view on the priority issue. However, it is apparent from his book what his opinion on this matter is. Weller's career did not end with his 1954 Nobel Prize. He made a number of seminal contributions after having received the award. He was the first to isolate the medically very important varicella, rubella and cytomegalo viruses. The first two viruses were recovered from his sons Peter (now a professor of medicine at Harvard Medical School) and Robert, respectively. The early and prompt recognition of Weller's and Robbin's contributions to the 1954 prize, together with the particular conditions of nomination of Theiler for his yellow fever work in 1951 provided the incentive for writing the next chapter describing other Nobel Prizes in Physiology or Medicine awarded under corresponding conditions.

Chapter 6

Unusual Nobel Prizes in Physiology or Medicine

Penicillin Vitamin K DDT
Corticosteroids Insulin

The procedures for selection of Nobel Prize recipients have evolved during more than a hundred years. Alfred Nobel's final will gave only limited guidance as to how the formal processes for this selection should be elaborated. Supplementary conditions were specified in the statutes of the prize-awarding institutions and the Nobel Foundation, established in 1900 to facilitate the management of the prizes. Still, even after these conditions had been defined, allowing the first prizes to be awarded, it took a long time until consolidated and uniform procedures for selection of prize recipients had evolved.

In the beginning there were extensive discussions at the Karolinska Institute about the interpretation of critical formulations in the will like "... , during *the preceding year*, shall have conferred the *greatest benefit on mankind*" and "... most important *discovery* within the domain of *physiology or medicine*." For the early prizes the citation was "... for (*or* in recognition of) his work on ... ," but from the 1919 prize in Physiology or Medicine (awarded in 1920) and onwards the formulation was, in most cases, changed to "... for his discovery of ... ."

At about that time there were some important changes in the management of the Nobel Committee. Originally it was chaired by the Vice-Chancellor (Rector) of the Institute, who was also responsible for the secretarial functions. In 1918 a half-time secretary was employed. This was Göran Liljestrand, who later became a professor of pharmacology. He remained secretary of the committee for 42 years (!) and obviously came to have an enormous influence on the development of the work of the committee[1,2]. The saying goes that, towards the end of his long term of responsibility, he formulated the protocols prior to the meetings of the committee, but that is probably just a good story.

Another influential person was a professor of physiology called Johan Erik (Jöns) Johansson. He had a major influence on Nobel when he formulated his final will (Chapter 1). Johansson became chairman of the Nobel Committee in 1918, a position he kept until 1926. In this responsibility he stressed that in the selection of prize recipients the main emphasis should be on scientific originality. The key word was *discovery*. The interpretation of the meaning of this term and what influence it may have on selection of prize recipients has been the subject of ongoing discussions by the committees until the present day. Some aspects of these deliberations were discussed in Chapter 4[3].

Göran Liljestrand (1886–1968) (right), professor of pharmacology and secretary of the Nobel Committee (1918–1960) and Jöns Johansson (1862–1938), professor of physiology and chairman of the committee (1918–1926). [Photo from Lennart Stjärne.]

Since 1943 (1944) there has been an uninterrupted sequence of prizes, but as there were nine years without any prizes before that, the prize in Physiology or Medicine in 2009 was the 100th one. General information on all these prizes and the laureates who have received them can be found at http://nobelprize.org. As a consequence of the 50 years secrecy rule that applies to Nobel archives material, the last prize that can be reviewed at the time of writing is the one given in 1959. Thus it is possible to give an overview of half of all the hundred prizes that have been awarded so far. A considerable amount of documentation on the advancement of different fields of biomedicine and the early prize recipients, and also frequently-nominated non-recipients representing these fields has been comprehensively reviewed by Liljestrand[2].

In previous studies of the 1954 and 1951 Nobel Prizes in Physiology or Medicine two unexpected observations were made. The 1954 prize[4] (Chapter 5) nominations were dominated by proposals for John F. Enders, who had already been recommended for a prize for three years. His two young collaborators Thomas Weller and Frederick Robbins were nominated for the first time in 1954, the same year in which they received the prize. This chapter presents other similar cases among all the prizes in Physiology or Medicine that were awarded between 1901 and 1959. In addition to the 1901 prize, which by

Table 1. Nobel Prizes in Physiology or Medicine awarded at the first year of nomination (awardees in parentheses not included).

Year	Name	Motivation
1901	von Behring, Emil A.	"for his work on serum therapy, especially its application against diphtheria, ..."
1912	Carrel, Alexis	"in recognition of his work on vascular suture and the transplantation of blood vessels and organs"
1922 (1923)	Hill, Archibald V. Meyerhof, Otto F.	"for his discovery relating to the production of heat in the muscle" (½ prize) "for his discovery of the fixed relationship between the consumption of oxygen and the metabolism of lactic acid in the muscle" (½ prize)
1923	Banting, Fredrick G. MacLeod, John J. R.	"for the discovery of insulin"
1948	Müller, Paul H.	"for his discovery of the high efficiency of DDT as a contact poison against several arthropods"
1950	(Kendall, Edward C.) Reichstein, Tadeus Hench, Philip, S.	"for their discoveries relating to the hormones of the adrenal cortex, their structure and biological effects"
1954	(Enders, John F.) Weller, Thomas H. Robbins, Frederick C.	"for their discovery of the ability of poliomyelitis viruses to grow in cultures of various types of tissue"
1958	Lederberg, Joshua	"for his discoveries concerning genetic recombination and the organization of the genetic material of bacteria" (½ prize)

definition had to be given in the same year as the first nomination, there have been, besides Weller and Robbins in 1954, a total of nine more recipients in the same category (Table 1). Apparently this situation was not as unusual as originally was conjectured. Figure 1 illustrates the interval between the first time of nomination and the time at which the prize was given for all the 64 laureates between 1911 and 1959. The 12 laureates of the first ten years were not included in the figure since it is not meaningful to measure this interval for the early prizes. The time interval varies extensively with an average of six years. It should be emphasized that the time of the first nomination is not the same as the time at which the discovery is made. In most cases there is a certain time-lag, sometimes a very long one, before the first nomination is made.

The 1951 prize to Max Theiler[3] (Chapter 4) was awarded in the absence of any external nomination. On the final day of nomination, January 31, 1951, Theiler was proposed by the chairman of the committee, with reference to the previous two external nominations submitted in 1948 and 1950. As can be seen in Table 2 there are four more cases of this kind: one, August Krogh, also nominated by the

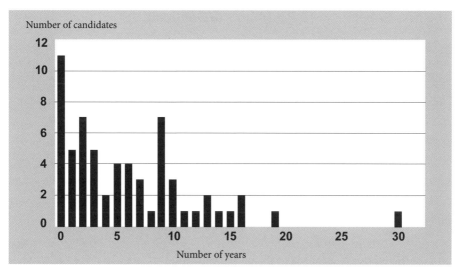

Figure 1. Time interval between the year of the first nomination and the year at which a prize in Physiology or Medicine was given (1911–1959).

chairman; one, Arthur Kornberg, nominated by a member of the committee; and two more cases, including three candidates, nominated by the secretary of the committee. It should be mentioned that it is a routine procedure since many years that the secretary may nominate candidates who have been runners-up during a preceding year, in case there are no external proposals.

In what follows the different — less — unusual Nobel Prizes in Physiology or Medicine, except the three cases discussed separately in Chapters 4, 5 and 7, will be presented in chronological order.

Table 2. Nobel Prizes in Physiology or Medicine awarded without external nomination (awardees in parentheses not included). Proposals were secured by a member of the Nobel Committee.

Year	Name	Motivation
1920	Krogh, August S.	"for his discovery of the capillary motor regulatory mechanism"
1943 (1944)	Dam, Henrik C. P. Doisy, Edward A.	"for his discovery of vitamin K" (½ prize) "for his discovery of the chemical nature of vitamin K" (½ prize)
1945	(Fleming, Alexander) Chain, Ernst B. (Florey, Howard W.)	"for the discovery of penicillin and its curative effect in various infectious diseases"
1951	Theiler, Max	"for his discoveries concerning yellow fever and how to combat it"
1959	(Ochoa, Severo) Kornberg, Arthur	"for their discovery of the mechanism in the biological synthesis of ribonucleic acid and deoxyribonucleic acid"

The first Nobel Prize in Physiology or Medicine

From its inception, the Nobel Prize in Physiology or Medicine was off to a solid start. Apparently the new prize attracted a good deal of attention. It was the first international prize and it was connected with a considerable monetary reward, since Nobel's original idea was to provide a 20-year scholarship. There were a total of 77 nominations for 44 candidates (Figure 2). The nominations included essentially all the front-line candidates at the time. All the recipients of the first eight years of Nobel Prizes in the field were proposed. A tradition of making thorough reviews by members of the committee or by other professors at the institute was started in the very first year.

There were extensive investigations, sometimes by more than one evaluator, of the work of Camillo Golgi, Santiago Ramón y Cajal, Ivan P. Pavlov, Niels R. Finsen, Emil A. von Behring, Ronald Ross and others. One report on Pavlov recapitulated a visit that members of the committee had made to his laboratory, where they had witnessed practical demonstrations of his experiments. A special review was thematically oriented towards the problems of malaria, comparing candidates who had made different contributions. Ross was evaluated to be the

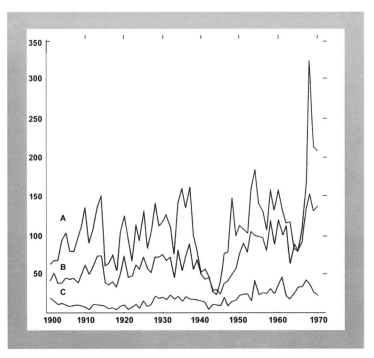

Figure 2. The number of nominations (A), candidates (B) and special reports (C) for prizes in Physiology or Medicine. Reproduced from Reference 2, p. 152.

strongest candidate in this important field and he received the 1902 prize in Physiology or Medicine "for his work on malaria, by which he has shown how it enters the organism … ." A few years later, in 1907, another prize in this field was given to Charles L. A. Laveran "in recognition of his work on the role played by protozoa in causing diseases".

Emil von Behring had received five nominations, and his discovery that a specific antiserum could prevent diphtheria was highly praised. It was concluded that he had priority in this finding and that the work of others was confirmatory. Immediately after von Behring's discovery, in 1895, there was a widespread use of specific serum therapy. This led a joint pair of reviewers to conclude:

Emil von Behring (1854–1917), recipient of the first Nobel Prize in Physiology or Medicine 1901. [From *Les Prix Nobel 1901.*]

> It is thus our view that both the fundamental discovery and the demonstration of its value for practical use have been made *so far back in time* that we, acknowledging that they are worthy of being recognized by a Nobel Prize, cannot propose them for a prize now.

Apparently the reviewers were thinking about the formulation in the will "… during the preceding year." The recommendation by the committee to the faculty (the College of Teachers) was instead to give one prize to Finsen and another to Ross. However, the faculty was of a different opinion and selected von Behring for the 1901 prize "for his work on serum therapy, especially its application against diphtheria, … ." In the end, the hesitation by the referees did not deter the professors at the institute from awarding him the prize. The question of how and if discoveries made some years back could be awarded remained a matter for discussion. A more comprehensive presentation of von Behring and the work that led to his development of serum therapy can be found in the book *Pioneers of Microbiology and the Nobel Prize* by Ulf Lagerkvist[5].

There came to be a dominance of prizes in the area of infections and immunity during the first 20 years of Nobel Prizes for Physiology or Medicine[2,6]. 8 out of 18 recipients were active in these fields. In 1903 the scientific giant Robert Koch, in whose laboratory von Behring had done his work, was passed over in favor of Niels R. Finsen[7]. His "contribution to the treatment of diseases, especially lupus vulgaris, with concentrated light radiation, …" lacked theoretical and scientific underpinnings, but was believed at the time to be of practical value. Finsen had followed an independent career and never wrote a medical dissertation. His international and national fame peaked in 1903 but he died less than a year later. Under the mounting pressure of nominations, Koch eventually received the prize in 1905 "for his investigations and discoveries in relation to tuberculosis."

One of von Behring's colleagues in Koch's laboratory was Paul Ehrlich. He was able to work out a reliable and reproducible standardization method for diphtheria serum. After this he went on to make many important discoveries and eventually established his own institute in Frankfurt. In 1908 he shared a Nobel Prize with Ilya I. Mechnikov (also Metchnikoff) "in recognition of their work on immunity." Ehrlich's unyielding and sometimes controversial personality led to conflicts with von Behring and the Swedish Nobel Laureate in Chemistry in 1903, Svante Arrhenius[8]. After he had received his prize, Ehrlich was again nominated, now for the "magic bullet" Salvarsan for the treatment of syphilis. This pioneering contribution to chemotherapeutic research, however, was not recognized by another prize.

The enigma of Alexis Carrel

The rapid advance toward a Nobel Prize

In 1912 Alexis Carrel, then at the Rockefeller Institute in New York, was nominated for the first and only time for a prize in Physiology or Medicine by the Paris-Sorbonne professor of general pathology and therapy, Charles Bouchard. This nomination first cited three other French scientists, including Charles Richet, who was to receive the prize the following year for his work on anaphylaxis. Carrel was recommended for a prize "for his work on the continuation of life in a latent and efficient form in isolated tissues with samples collected both from cadavers and living people and conserved in coagulated liquid." This part of Carrel's work was *not* the main focus of the assessment performed by Jules Åkerman, professor of surgery at the Karolinska Institute. His 26-page long evaluation mainly concerns Carrel's work on vascular anastomosis and organ transplantation with a brief allusion to Carrel's more recent work on cultivation of tissues under laboratory conditions.

Åkerman gave a brief history of methods of making vascular anastomosis. Then he presented Carrel's large number of publications, one after the other. It started with the work performed in Lyon using a triangulated form of suture using fine needles and as advanced antiseptic conditions as possible. Hereafter Åkerman discussed how Carrel used his technique to transplant whole organs or parts of organs, like the thyroid gland, the spleen, the kidney, the ovary, etc. It was noted that transplantation within the same animal was possible, provided concurrent infections did not occur, but that transplantation between animals was not successful, although Carrel sometimes claimed this. The reasons for rejection of transplanted tissues were unknown at the time. The main part of Åkerman's long review describes the content of some 30 of Carrel's publications in turn. Only a very few judgements on their significance and mutual relationships were given. It was mentioned that organs kept in the cold for some time had been successfully transplanted. It was speculated what use the techniques developed may have for humans.

Five of the publications discussed in the report describe Carrel's attempts to culture cells outside the body in laboratory vessels. In these studies he used a technique introduced some years earlier by Ross G. Harrison. Carrel could grow normal cells from dogs, cats, hens and other species and also cells from tumors. The cells were shown to divide and subcultures were made. The developments of the cultures could be observed by the light microscope. Carrel made some technical improvements and showed that cells could be kept alive for many months provided that no microbial contamination occurred. In studies after 1912 Carrel became more involved in the cell culture studies, but until then only qualitative results of limited news value were obtained.

In the last paragraphs of Åkerman's investigation, he noted that the committee had requested him to evaluate Carrel's work on tissue and organ transplantation and surprisingly not for the work for which he was nominated by Bouchard. Åkerman stated, however, that on his own initiative he wished briefly to mention the tissue culture work. He noted that this work was a follow-up to work done by others and did not allow Carrel to claim any priority. Carrel's contributions to organ transplantation were praised, however, as being of great theoretical interest. Potentially they may in the future had importance also for man. Åkerman's concluding sentence stated: "According to my opinion the question of awarding a Nobel Prize to Carrel should be given serious consideration."

In 1912 the committee had subjected the contributions of ten other scientists apart from Carrel to full reviews. The committee's unanimous conclusion was that the prize for 1912 should be awarded to Carrel "for his work concerning vascular suture, transplantation of blood vessels and organs and the culturing

Alexis Carrel (1873–1944), recipient of the Nobel Prize in Physiology or Medicine 1912. [From *Les Prix Nobel 1912*.]

and growth of tissues outside the organism." This was also the decision of the faculty, but they crossed out the final part of the motivation, i.e. the only part mentioned in Bouchard's single nomination.

Consequently the introductory speech by Åkerman at the Nobel Prize ceremony, the first one given by a specialist in the field (in surgery), only referred to Carrel's contributions in the field of surgery[9] and so did Carrel's Nobel lecture[10]. Carrel's award was a truly applied clinical prize and his contribution had some long-lasting impact[11]. Carrel was the first US citizen to receive a Nobel Prize in Physiology or Medicine, and he was also the youngest prize recipient to date. He was then only 39 years old.

Carrel's position as a senior scientist at the prestigious Rockefeller Institute, led by the highly influential scientist Simon Flexner (Chapters 4, 5), and as a Nobel Prize Laureate, gave him a unique platform for future endeavors. His life came to take many turns with many colorful and controversial encounters. It included the material which has ensured that books and presentations of his life keep appearing even into the present century. In the book titles he is referred to as "visionary surgeon"[12], "God's eugenist"[13], "The Immortalists"[14] and other imaginary descriptions[15]. Clearly Carrel was not a person to leave people he encountered in his milieu unaffected.

Eternal life in cultures

Carrel's major engagement after the prize was his tissue culture work. This field had been pioneered by Harrison, who developed methods of growing embryonic tissue from tadpoles in coagulated lymph. He made important observations on embryo differentiation in cultures. Harrison was nominated 19 times for a prize between 1913 and 1941. In 1917 the committee recommended him for the prize, but no prize was given in that war year. In the end Harrison did not receive a prize, but the field was eventually recognized in 1935, when Hans Spemann received a prize "for his discovery of the organizer effect in embryonic development."

Carrel's tissue culture work had a different focus. He argued that he had managed to establish the eternal life of cells outside the body[16]. Particularly spectacular were his cultures of embryonic chicken heart cells which continued to beat in the laboratory vessels for more than 100 days. Eventually they stopped beating but continued to live on, according to Carrel, in eternity. However, the claims that cells could be maintained in repeated consecutive cultivations for over ten years were flawed. Many years later it was discovered that normal, non-cancerous cells have a limited capacity for repeated cultivations. The problem with Carrel's cultures was that they were not amenable to quantitative biological assessments. They might have been tested for studies of the production of biologically active components in the test tube or for the growth of viruses, but Carrel did not take his studies in these directions.

He did in fact, however, perform one experiment together with another exceptional and highly profiled Rockefeller Institute personality, Dr Hideyo Noguchi[17] (Chapter 4). In parentheses it can be mentioned that the latter scientist probably holds one of the records when it comes to non-reproducible "discoveries." He was nominated many times for a Nobel Prize, but the committees judged his work correctly and never awarded him a prize. The Carrel–Noguchi experiment was meant to demonstrate production of antibodies in the laboratory vessels, but it cannot be concluded whether the hemolysis observed in their test system was due to antibodies or to something else. The tests were not repeated, and much later it remained for others to demonstrate the production of antibodies in tissue cultures. It was only in 1984 that Georges J. F. Köhler and Cesar Milstein were awarded a prize for "the discovery of the principle for production of monoclonal antibodies."

Carrel's engagement during the two world wars

Carrel spent the time of the First World War in France attempting to introduce new techniques for antiseptics. He then returned to the Rockefeller Institute and his tissue culture work. Gradually he became more and more involved in philosophical and political issues. He was influenced by the inter-bellum pessimism and adopted a holistic view into which he mixed eugenics and spiritualism. He made apocalyptic warnings about the decline of the white race. In the late 1930s he published his best-selling book *Man the Unknown* — L'homme cet inconnu[18] —that made him the intellectual icon of the conservative right. The book sold more than 400,000 copies and was translated into 14 languages.

It was Carrel's ambition to establish an Institute of Man in the US. In these endeavors he was supported by none other than the world-renowned aviator

Alexis Carrel (right) and Charles Lindbergh. [Photo from Reference 13.]

Charles Lindbergh. During the late 1930s, they developed a friendship based on mutual admiration and shared political views. When Lindbergh visited Germany in the winter of 1938, he met with Hitler and Göring. This visit was much criticized in the US, but Carrel, who in his book had praised Hitler and Mussolini, defended it. In 1939 Carrel, to his great disappointment, was forced to retire by the Rockefeller Institute's new director Herbert Gasser, then forthcoming 1944 Nobel Laureate in Physiology or Medicine. In contrast to the previous director Simon Flexner, who remained a supporter of Carrel throughout his life, Gasser was critical of Carrel's metaphysical and non-scientific argumentations. Carrel then returned to France and developed a close collaboration with the Vichy government. He received support to start developing an Institute of Man, something that he had always dreamt of. In 1944, as the tide of war was changing, Carrel was suspended from his association with the institute and in November of that year he died.

Carrel was never actually accused of any war crimes and there is no evidence that his enthusiasm for eugenics had any practical consequences. To a certain extent they have to be viewed in the historical perspective of their time. As a curious fact it can be mentioned that the first proposal in 1918 for a Nobel Institute at the Karolinska Institute was an Institute for Heredity and Racial Biology. It was rejected by a narrow majority of the faculty. Carrel's reputation in the US was heavily tarnished by his association with the Vichy government, and it took time for it to be restored, and then only in part. In 1973, on his centenary, streets in France were called after him and a few years earlier the University of Lyon had named one of its medical schools after him. This, however, did not last more than a few decades, and Carrel was again attacked as a racist and his name removed from the medical school.

Did Carrel's scientific contributions motivate his Nobel Prize in 1912?

His tissue culture work for which he was nominated certainly did not. The committee, however, came to focus on his contributions as a surgeon and not as a non-quantitative biologist. His introduction of an improved technique for vascular surgery was of importance for future transplantation surgery and hence of benefit to mankind. But as the criteria for Nobel Prizes in Physiology or Medicine came to evolve with a particular emphasis on *discovery,* he would not, had he been evaluated by later committees, have become a recipient of the prize. His contribution can be described at best as an *improvement,* which can be applied to motivate a prize in Chemistry, but not in Physiology or Medicine. The fundamental concepts for transplantation immunology were developed first in the 1950s and were awarded in 1960 with a prize to F. Macfarlane Burnet (Chapter 3) and Peter B. Medawar "for discovery of acquired immunological tolerance." The introduction of techniques to successfully use transplantation for human health care came much later still and this was recognized by the award of the 1990 prize to Joseph E. Murray and Thomas E. Donnall "for their discoveries concerning organ and cell transplantation in the treatment of human disease."

Overall it seems that committees over the years have not been very enthusiastic about contributions in the field of surgery. This is perhaps because it is difficult to identify discoveries in this area. The prize awarded to Emil T. Kocher in 1909, for his work on the thyroid gland, mentions surgery, but later famous surgeons never became strong competitors for the prize. One example is Ferdinand Sauerbruch in Zürich, who received 56 nominations between 1914 and 1951, and others are Alfred Blalock and Helen B. Taussig, at Johns Hopkins University, Baltimore, Maryland, pioneers in heart surgery, in particular treatment of the blue baby syndrome, who also received many nominations. If the award of a prize to the latter candidates, perhaps including Vivien T. Thomas, who was never nominated, had materialised, it would have improved the poor records of both female Nobel Laureates and of the complete absence of African–Americans among the prize recipients. The closest to a follow-up prize to the one Carrel received is the 1956 prize awarded to André F. Cournand, Werner Forssmann and Dickinson Richards "for their discoveries concerning heart catheterization and pathological changes in the circulatory system."

The importance of friendship across the Kattegat

The Nobel Prize in Physiology or Medicine awarded in 1920 to S. August S. Krogh "for his discovery of the capillary motor regulating mechanism" has been thoroughly reviewed[19]. It was based on three nominations, two in 1919 and one in 1920. One of the 1919 nominations was from a Danish colleague and collaborator, the professor of gymnastics theory J. Lindhard, the other from N. Zuntz, professor of zoophysiology in Berlin. Their motivations were rather general, like "publications of numerous works on the supply of oxygen in the organism," and did not focus on Krogh's most recent discoveries regarding capillary functions. Zuntz's nomination included two other physiologists focusing on metabolism, Joseph Barcroft from Cambridge and Francis G. Benedict from Boston.

Krogh and Barcroft were subjected to two separate full evaluations by the chairman of the committee, Johansson. His preliminary two-page report mostly discussed the capillary studies, which had been published in three

articles in Danish in 1918 and in English in 1919, the year of the first nominations. His conclusion was that the results should be confirmed and that the contribution therefore did not qualify for a prize. In spite of this 'wait and see' attitude the committee wanted to have a further analysis of Krogh's work and Johansson therefore wrote a full, seven-page, report. The report compared the two major theories of oxygen distribution: secretion proposed by the renowned physiologist Christian Bohr, Krogh's mentor and Niels Bohr's father, and diffusion, as favored by Krogh. It then presented

August Krogh (1874–1949), recipient of the Nobel Prize in Physiology or Medicine 1920. [From *Les Prix Nobel 1920.*]

how Krogh had attempted to prove the correctness of his theory experimentally, and again noted that certain questions had to be settled before the theory could be seen as validated. Still, this

second report was somewhat more positive to Krogh's candidature. Two out of the six members of the committee in fact argued in favor of giving the prize to Krogh as early as 1919, but the faculty decided to hold back the prize that year. In 1920, there were no external nominations for Krogh, so a nomination was made by none other than Johansson, with reference to his report the previous year.

In 1920, Johansson and another member of the committee, John Sjöqvist, professor of chemistry and pharmacy, paid a visit to Krogh's laboratory. Their experiences on this visit became the essence of the second full report that Johansson wrote about Krogh's candidature. The report provided no additional support for Krogh's theory and concluded: "However, as it is generally believed that Krogh will sooner or later win general recognition for this theory, and since he is undeniably an attractive Nobel Prize candidate in all respects, it is easily comprehensible that already at this stage, I lean towards recommending a prize."

The committee was quite divided in its recommendations. A majority, including the committee's powerful chairman and the only reviewer used for Krogh, was in favor of the reserved 1919 prize being given to him, and a minority favored giving a prize to Karl Neuberg, professor in Berlin, for his work on fermentation and cleavage of sugar. On this occasion the faculty took control of events and selected Jules Bordet "for his discoveries relating to immunity" for the 1919 prize and Krogh for the 1920 prize. Whereas Krogh had two external nominations, Bordet had received 106 nominations since 1902. He had been evaluated repeatedly and considered prize-worthy.

Because of his personal contacts with both the chairman and the secretary of the Nobel Committee, Krogh's Nobel Prize has been referred to as inevitable[20]. He was the second non-medical scientist to receive a Nobel Prize in Physiology or Medicine, the first one being the Russian Ilya Mechnikov, who shared it with Ehrlich in 1908. Since Krogh's critical articles, on which the award was based, were published in English in 1919, it can be said that for once the committee managed to fulfil the conditions in Nobel's will that stipulated "… to those who, during the preceding year, shall have … ."

Krogh's Nobel fame brought him into wider contact with the world of science. In 1922 he was invited to undertake a lecture tour in the US and visited a number of famous universities. The topic of the day was the new wonder drug insulin, and in order to learn more about it and since Krogh's wife Marie was a diabetic, he contacted the Scottish physiologist John MacLeod in Toronto and arranged a visit. This visit had important consequences, as will become apparent later.

The attraction of contraction

Powerful developments in a subfield of physiology

In 1922 the Nobel Committee decided to subject only six candidates to full reviews. Three of these were later to receive Nobel Prizes in Physiology or Medicine: J. Fibiger, who was awarded the prize in 1926 for a discovery that could never be confirmed[21,22], E. C. Kendall, who was nominated for his studies of the hormone thyroxin, but eventually, as will be discussed below, received a prize in 1950 for his work on adrenal hormones, and Karl

Landsteiner, who received a prize for the discovery of blood groups in 1930. However, in 1922 the committee decided that none of the proposed candidates deserved to receive the prize and recommended that the money be allocated to a special fund, its use to be reconsidered in 1923. The faculty agreed with this.

In 1923 the committee took a more positive stand on the proposed candidates and recommended recipients for both the 1922 and the 1923 prize. For the 1922 prize recent discoveries in muscle physiology caught the committee's interest. Three candidates, Gustaf Embden, Archibald V. Hill and Otto F. Meyerhof, were proposed. Of these three, Embden received three nominations. One of these was by K. Hürtle from Breslau and it also recommended Meyerhof. Another all-encompassing proposal was by professor A. Bethe from Frankfurt, who nominated all three scientists, under the subheading of muscle physiology. The third nomination by E. Schmitz, also from Breslau, only commended Embden. The Nobel Committee decided to conduct two full investigations, one of Hill and Meyerhof by professor Johansson, and a separate one of Embden by another of its members, John Sjöqvist, professor of chemistry and pharmacy.

Johansson's thorough review of Hill and Meyerhof was summarized by noting that there has been a "colossal" development in the field of muscle physiology in the last decade and that no one can deny that it deserved recognition by a Nobel Prize. He also stated that at the time no other biological process is as well elucidated as the contraction process of muscles. The last two paragraphs of his evaluation read:

> If the three researchers investigated had worked together, the formulation of a proposal for the awarding of the prize this year would be very simple: a prize for the discovery of the physical chemical processes behind the muscle process should be awarded jointly to professors Embden, Hill and Meyerhof, who together have made this discovery. However, the situation is complicated by the fact that the three researchers have worked independently, and only the general flow of thoughts and in part the research methodology have been shared. In fact it would be difficult to imagine that such widely encompassing and, as regards resources, highly demanding work, could have been concentrated in a single place. The question one can then pose is: are the separate contributions, which each one of the three researchers has made to the joint effort, of such a nature that they can be viewed as representing a single discovery, and

Archibald Hill (1886–1977) and Otto Meyerhof (1884–1951), recipients of the Nobel Prize in Physiology or Medicine 1922. [From *Les Prix Nobel 1922.*]

can one attribute to this such a degree of importance that it is necessary to award a Nobel Prize?

It can be deduced from the investigation presented above that the most important contribution by Hill is his discovery concerning the adjustment of heating in the activated muscle; by Meyerhof, the discovery of the fixed relationship between consumption of oxygen and the turnover of lactic acid in the muscle, and by Embden the discovery of the lactacidogen. I find it obvious that each one of these discoveries has the importance that the statutes require for the awarding of a prize. Like the proposer Bethe I find it impossible to put any of these contributions ahead of the other.

In the separate evaluation by Sjöqvist, Embden's comprehensive contributions to the field of muscle physiology were reviewed in a respectful and appreciative tone. He concluded:

It is apparent from this presentation that I have a high regard for Embden's work. As a final judgement I would like to stress that I consider his discovery and preparation of lactacidogen, investigations of its biological significance and discoveries regarding the importance of muscle permeability for the contraction of muscles to be highly deserving of (being awarded) a Nobel Prize in Physiology or Medicine.

However, something important must have happened in the deliberations by the committee, because its final recommendation to the faculty for the 1922 prize mentions only Hill and Meyerhof, and Embden is excluded. Most likely, Hill's and Meyerhof's contributions were considered more fundamental. In fact the history of research on muscle physiology supports this conclusion[23] (Bengt Saltin, personal communication). One other factor that might have been of importance is that until this time no Nobel Prize in Physiology or Medicine had been given to more than two persons. Maybe the committee was still reluctant to start bestowing a prize on three people, and Liljestrand[2], the secretary of the committee, noted that there were statutory restrictions at the time, which prohibited awarding this many candidates. The first time a prize was shared equally between three persons was in 1934 when George H. Hoyt, George R. Minot and William P. Murphy were rewarded "for their discoveries concerning liver therapy in cases of anemia," based on the identification and characterization of vitamin B12.

When the faculty met on October 11 they accepted the Hill-Meyerhof nomination without hesitation. Both laureates happily came to Stockholm and gave their Nobel lectures two days after the prize ceremony on December 10. The laudatory speech by Johansson[24] highlights the nationality of the awardees and the aftermath of World War I, and notes that the joint recognition "… gives a clear expression of one of the ideas upon which the will of Alfred Nobel was founded, that is, the conception that the greatest cultural advances are independent of the splitting up of mankind into contending nations."

Embden never received a Nobel Prize

During the subsequent nine years, 1924–1933, Embden was nominated eight more times and he was subjected to four more full investigations. The last one was in 1932 by Einar Hammarsten, professor of medical chemistry. It concluded:

> By the establishment, during the last two years, of the final evidence that the metabolism of organic phosphorous compounds is intercalated between the formation of lactic acid and the contractile moment, Embden's discovery of the essential importance of organic phosphoric acid compounds for the contraction of muscles has further gained in clarity and importance. The opinion, which I have expressed in my investigations of 1929 and 1930, that this discovery is prize-worthy, has therefore been further strengthened.

This conclusion was echoed in the final summary and recommendations by the committee to the faculty, but Embden did not make it to the short list. This was Embden's last opportunity; he died in 1933. Embden remains in the history books for his description of the cellular metabolic sequence from glycogen to lactic acid known as the Embden–Meyerhof pathway.

Instead of Embden the committee prioritized Charles S. Sherrington and Edgar D. Adrian, and this choice became the decision of the faculty. The motivation was "for their discoveries regarding the functions of neurons." Sherrington had to wait 30 years for his prize (Figure 1). During this time he accumulated 134 nominations from persons representing 13 countries. He was 75 years old, one of the two oldest recipients of the prize between 1901 and 1959 (see figure on page 31).

Insulin, its remarkable discovery and a much debated Nobel Prize

A committee in quandary

The report by the committee to the faculty in 1923 contained a second proposal, to give the 1923 prize to Frederick G. Banting and John J. R. MacLeod. This proposal was not as well received by the faculty at the October 11 meeting as the recommendation for the 1922 prize. It was returned to the committee for reconsideration. What was the problem?

The discovery of insulin is one of the most thoroughly reviewed events in human medicine. In particular the carefully researched material used in the preparation of the book by Michael Bliss[25] provides information to which little can be added. Still it is tempting to revisit the events. The discovery was spectacular and it is fully understandable that the committee took this opportunity, one of the few offered, to fulfill all criteria for the prize including that it shall be given to the person "who, during the preceding year, shall have conferred the greatest benefit on mankind." Nobel's original idea was to identify a promising young scientist who had made a brilliant discovery and who, by receiving the prize, would not have to think about any salary for the coming 20 years. Banting and his discovery fulfil all these criteria. He was only 32 years old when he received the prize in Physiology or Medicine, and he remains today the youngest person ever to have received this prize (Figure 4, Chapter 1). So, the prize awarded to Banting is perfect, had it not been for the fact that the head of the laboratory, John MacLeod, was also included. Let me briefly review how this came about.

There were four nominations for the discovery of insulin, two of them for Banting alone. One came from G. W. Crile, a distinguished professor of surgery in Cleveland and the other from the Harvard professor Francis G. Benedict. The latter recommendation was very strong and emphasized repeatedly that it was Banting, and no one else, who had priority in the discovery. There was one nomination for both Banting and MacLeod by the recent Nobel Laureate August Krogh, who later came to have considerable influence on the committee. It was he who argued that MacLeod's overriding responsibility for the laboratory was indispensable for the work that Banting did. There was also one nomination for MacLeod alone, from a good friend of his at Case Western Reserve University in Cleveland.

Having received these nominations, the committee decided to let two of its members, Sjöqvist and H. Christian Jacobaeus, professor of medicine, make full investigations of the two candidates. Sjöqvist gave some background to the discovery and then highlighted how Banting got his brilliant idea to close the connection between the pancreatic gland and the intestine and let the cells responsible for producing enzymes for digestion of food degenerate. He hoped, correctly as it turned out, that the cells in the Langerhans' islands in the gland producing hormone would survive. It was from such material that crude insulin (a name for the postulated hormone conferred as early as 1916 by Sharpey Schafer) could be extracted without being destroyed by the proteolytic enzymes of the gland. Banting took his idea to MacLeod, who according to Sjöqvist "showed a lively interest." Others have expressed it differently. MacLeod was in fact absent from the laboratory to visit his homeland of Scotland during the summer months when the critical experiments were performed. As a consequence he was not a co-author of the critical publications eventually leading to the first successful treatment of a 14-year old, severely ill diabetic boy between January and February 1922. It was *after* that finding that MacLeod reoriented the whole laboratory to work on insulin. Until then he had provided Banting with two young assistants, Charles Best and later James B. Collip for the chemical work, and also with the laboratory facilities required for their work.

Just before discussing the first successful clinical application of insulin Sjöqvist wrote:

At the Eleventh International Congress of Physiology in Edinburgh at the end of July this year, at which the insulin problem, after an introductory lecture by MacLeod, was the topic of discussion, Banting announced that he had managed to demonstrate that insulin normally occurs in the blood of healthy individuals, dogs and rabbits, but in contrast not in the blood of dogs from which the pancreas has been removed, an observation of the uttermost importance.

The two last paragraphs of Sjöqvist's investigation ran:

It is my personal opinion that the discovery of insulin in the pancreas, studies of its physiological effects and its introduction into therapy is of the scope that *it can be rewarded* with a Nobel Prize in Physiology or Medicine.

When it comes to the possible awarding of the prize and its distribution I side with the nominator (Krogh), who is of the opinion that it should be given jointly to Banting and MacLeod. The honor of the idea and initiative goes to Banting. As can be judged from the publications available and *assertions from associated persons,* MacLeod, at whose department the investigations have been conducted, has been the leader of the scientific investigations, and it seems without doubt that it is thanks to his great contribution that the discovery has taken on the importance it now has. It deserves to be mentioned that it certainly was not an accident that Banting took his idea to MacLeod, since he had already previously performed important investigations of carbohydrate metabolism.

It was Jacobaeus's responsibility to supplement Sjöqvist's evaluation of the physiological role of insulin, to evaluate its practical importance. He described the explosive spread of the clinical use of the hormone and the applications at different stages of diabetes. In order to convince himself and his colleagues in the committee about the importance of the discovery of insulin and its use, Jacobaeus had requested witness statements from some leading diabetologists. Their answers cover about half of his 10-page review. Based on the comments received, he became convinced of the documented, fundamental importance of insulin for the treatment of diabetes. He therefore stated that: "It is my unhesitating view that the discovery of insulin can be awarded a Nobel Prize already this year." He then discussed who should receive the award and noted that there was no doubt about the fact that it was Banting who had the original idea and who first performed the experiments. He "should therefore be awarded the prize *in the first hand.*" He then continued,

On the other hand it is more difficult to decide on MacLeod's contribution, since it is not apparent from the literature. MacLeod, who is the head of the physiological laboratory in Toronto, has previously carried out investigations of blood sugar. Banting came to MacLeod with his idea and developed insulin under his leadership. I *have been told* that it is very likely that the discovery would not have been made if MacLeod had not supervised him, at least not as rapidly as is now the case. Banting *is said* to have even considered an experimental arrangement, which would not have led to the goal, something that was corrected by MacLeod. The question therefore is whether Banting alone shall be awarded the prize or if this should be awarded jointly to him and MacLeod. On the basis of what I have presented above I am most prone to give Banting and MacLeod a joint Nobel Prize.

In addition to these full reports the committee had available a special report from its secretary, who had been present during the insulin sessions at the congress in Edinburgh. Based on these thorough documents the committee decided to recommend a joint prize to Banting and MacLeod. As mentioned above, this was not well received by the faculty at its October 11 meeting. After discussions the recommendation was sent back to the committee for reconsideration. The main objections were formulated in a letter to the committee by a faculty member, Alfred Pettersson, a professor of bacteriology. In this letter he stated:

> During the time I have participated in the awarding of the Nobel Prize, the justification for the award has never been based on hearsay evidence from unknown persons, on statements like "it is beyond doubt," on things that are thought of as "very possible." In my opinion it is very necessary that the faculty adhere only to verifiable facts. Otherwise the faculty risks development of unpleasant disclosures at a later date.

Hereafter he discussed the contradictions in the concluding paragraphs of Jacobaeus's evaluation.

The committee took the critique seriously and made an extensive written statement. It refers both to the events at the Edinburgh meeting, where the committee had been represented, and to the contacts with Krogh, who was the most important source of information about the role of different individuals in the Toronto laboratory. The statement concludes, "… it is not possible to make a more thorough investigation of this discovery and the relative contributions

Frederick Banting (1891–1941) and John MacLeod (1876–1935), recipients of the Nobel Prize in Physiology or Medicine 1923. [From *Les Prix Nobel 1924–1925*.]

by Banting and MacLeod, nor is it necessary." The final decision of the faculty to select Banting and MacLeod was taken on October 18. In parentheses it can be added that because of this delay in announcing the prize three new nominations for the 1924 prize had already been submitted, posted on October 17–22. All of them were for Banting only.

The awardees' reaction to the prize announcement

When the announcement reached Banting on October 26 he was furious. According to his own words: "I rushed out and drove as fast as possible to the laboratory. I was going to tell MacLeod what I thought of him. When I arrived at the building, Fitzgerald (the director of the Connaught Laboratories; my remark) was on the steps. He came to meet me and knowing that I was furious he took me by the arm. I told him that I would not accept the prize; that I was going to cable Stockholm and not only would I not accept, but that they and the old foggy Krogh could go to hell. I defied Fitzgerald to name one idea in the whole research from the beginning to the end that had originated in MacLeod's brain — or to name one experiment that he had done with his own hands. Fitzgerald had no chance to talk … ."

Banting eventually calmed down and was persuaded to consider both the honor of being the first Canadian to receive the prize and the value of the prize for the branch of science concerned. As part of accepting the prize he also declared that he would share the money and the credit with Best, his closest collaborator, the then 24-year-old student that MacLeod had assigned him two years earlier.

The news about the prize reached MacLeod a few days later on November 2 when he landed in Montreal on a boat from England. He was very restrained in his comments. After he had heard that Banting intended to share his half of the prize with Best, he decided, to give fairness to the science efforts, to share his half with Collip. His knowledge of chemistry was essential to the purification of insulin that could be effectively used in humans. As a consequence the financial reward as a part of the prize was eventually divided among four persons, whereas the Nobel statutes specify that a maximum of three persons can share a prize.

Neither Banting nor MacLeod came to Stockholm for the prize ceremony on December 10. Sjöqvist gave the laudatory speech[26] and in their absence the British Minister received Banting's and MacLeod's prizes from the hands of His Majesty the King. There is no hint about any controversy in Sjöqvist's presentation. It states that "… Frederick G. Banting conceived an idea that was to prove of extraordinary importance for its future development," and later on "He imparted this idea to professor MacLeod of Toronto, after which, together with several fellow-workers, among whom I should like to mention especially Best and Collip, he began to work under MacLeod's guidance and in his laboratory in May 1921."

MacLeod delivered his Nobel lecture on May 26, 1925 in Stockholm[27]. It provides a very comprehensive account, including events both before and after the discovery. Only one paragraph is devoted to the discovery itself, where it reads "… F. G. Banting suggested preparing them from duct-ligated pancreas, and with the aid of C. H. Best and under my direction, he succeeded in 1921 in showing that such extracts reduced the hyperglycemia and glycosuria in depancreatised dogs." And further on "… the time seemed ripe to investigate their action on the clinical form of diabetes. This was done by Banting in a severe case under the care of W. R. Campbell, with the result that the hyperglycemia and glycosuria were diminished."

Banting gave his Nobel lecture in Stockholm a few months later, on September 15[28]. He presented the origin of his revolutionary idea and then went on to say: "On April 14th 1921, I began working on this idea in the Physiological Laboratory of the University of Toronto. Professor MacLeod allotted me Dr Charles Best as an associate." The only other place when MacLeod is mentioned in the comprehensive lecture (20 printed pages) is after a paragraph discussing the fact that the high protein content rendered continuous use undesirable, due to formation of sterile abscesses. There the text reads, "At this stage in the investigation, February 1922, Professor MacLeod abandoned his work on anoxemia and turned his whole laboratory staff on the investigation of the physiological properties of what now is known as insulin." Hereafter he introduced the biochemical purification procedure developed by Collip.

The post-prize developments of the two laureates and two non-laureates

The Nobel Prize provided Banting with a unique platform for future research. However he never came up with any additional major original ideas or experimental innovations. And never again was there such a unique frenzy of pioneering research as the discovery of insulin caused during the years 1921 to 1923. Banting mellowed during the 1930s, but he never became reconciled with MacLeod, who in 1928 left Toronto a disappointed man to return to Aberdeen. He never talked about the Toronto days and died in 1935 at the age of 59. During the Second World War, Banting was appointed coordinating chairman of Canada's wartime medical research efforts, which brought him to London. On a return visit he had dinner with Collip in Montreal on February 19, 1941, and then took off for Gander to continue to London for a second time. His plane, a Hudson bomber, crashed in Newfoundland and that was the end of Banting's eventful life.

Best and Collip went on to do good quality biomedical research. Best succeeded MacLeod as professor of physiology and Collip had a productive career as chairman of biochemistry at McGill University. Both were repeatedly nominated for a Nobel Prize in Physiology or Medicine for their work. Best was

Frederick Banting, Charles Best, and James Collip. [Photo from Reference 25.]

nominated 14 times between 1950 and 1954 and he was subjected to four separate full investigations. In the 1950 evaluation, Ulf von Euler, professor of physiology, concluded that Best's later discovery of the lipotropic effect of choline was prize-worthy. von Euler's evaluation includes an interesting appendix, discussing Best's role in the discovery of insulin, which I will return to below. The committee agreed with von Euler that Best should be declared prize-worthy for his work on choline. This conclusion appeared repeatedly in evaluations and committee protocols in 1951, 1952 and 1954, but Best was never to receive his own prize. Collin was nominated 11 times between 1928 and 1956 and four full investigations were made, the last one in 1951 by Hammarsten. None of them found Collip's contributions prize-worthy. These deliberations will be further discussed below.

The full intensity of the conflicts between the scientists involved never fully reached the press. At that time the *Toronto Star* had an overworked and weary journalist who wrote to make a living. His name was Ernest Hemingway, the recipient of the Nobel Prize for Literature in 1954. Had he appreciated the rich source of insights into human emotions that the conflicts between scientists may provide, he could have used this for inspiration, but he came to choose other settings for his examination of the turmoil of human ventures. However, a previous fellow laureate, Sinclair Lewis, Nobel Prize for Literature in 1930, did examine the world of scientific endeavors in his book *Arrowsmith*, published in 1925. Many of the locations and characters in the book are thought to be based on the Rockefeller Institute in New York. The lead character, Max Gottlieb, is modeled on Jacques Loeb, a world-famous scientist studying artificial parthenogenesis. He was nominated 79 times between 1901 and his death in 1924, but never received the award. In the novel Lewis fictionalizes the application of bacteriophages as a therapeutic agent. Incidentally it was in 1926 that d'Herelle, one of the discoverers of this kind of virus, was proposed by the committee for a prize he never got (Chapter 3).

Would the committee have changed its stand if it had delayed the awarding of the prize?

This is highly unlikely considering its elaborate written statement to the faculty. An important source of information was the 1920 Nobel Laureate Krogh, a good friend of both the chairman and the secretary of the committee. The story of how Krogh's interest in insulin arose deserves to be recapitulated.

During his lecture tour in the US in 1922, where he was accompanied by his wife Marie, he met the famous diabetologist Eliot P. Joslin, who told him that insulin had been discovered and purified in Toronto. This information was of particular importance to the family, since Marie had a recently-discovered, late onset diabetes. In addition Krogh was a good friend of the leading diabetes researcher in Denmark, Dr Hans C. Hagedorn. It was then arranged for Krogh to visit the Toronto laboratory, where he was kindly received by MacLeod, his host for three days in late November. He probably never met Banting. It was based on these impressions that he gave the firm advice to the Nobel Committee to award the prize to both Banting and MacLeod. But Krogh still had other fishes to fry. He became the father of what is referred to as the Danish "insulin adventure." Krogh managed to obtain exclusive rights to the production of the hormone in Scandinavia. It is assumed that Banting was not involved in these negotiations. The initiative led to the establishment in 1923 of the *Nordisk Insulin Laboratorium*, headed by Hagedorn and with Krogh as one of the three board members.

Finally, at the time when the discovery was made, the department chairmen of academic institutions exerted a decisive influence and controlled all the resources. Young scientists could not apply for independent scholarships and for research grants, and for Banting, who had no PhD, it was necessary to find a mentor and a patron. This was a "Herr Professor"-situation that the professors at the Karolinska Institute could identify with.

An exceptional comment by a later committee on the insulin prize

In Bliss's thoroughly researched book, the conclusion is that the combination of Banting and MacLeod was probably the most reasonable decision at the time. However, Bliss's book was published in 1982, which means that he could only examine the Nobel archives up to 1931, not those containing the 1950 archives. Still, this material is highly relevant, since, for once, a later committee commented on the decision of an earlier committee. One of the three nominations for Best in 1950 was none other than the famous neurophysiologist Sir Henry H. Dale, who himself had received a Nobel Prize in Physiology or Medicine in 1936. One part of his nomination under the heading *Other Scientific Achievements* deserves to be cited in full:

C. H. Best, then a young man who had not yet completed his medical graduation, shared with the late F. G. Banting the discovery of insulin in

Henry Dale (1875–1968) and Ulf von Euler (1905–1983), recipients of shared Nobel Prizes in 1936 and 1970, respectively. von Euler was chairman of the board of the Nobel Foundation (1965–1975). [From *Les Prix Nobel 1936, 1970.*]

1922. The work was made in the laboratory of the late J. J. R. MacLeod, of whom Best had, till then, been the pupil. It was MacLeod indeed who persuaded Banting that the research which the latter was determined to undertake needed the collaboration of somebody like Best, with recent knowledge of and training in the determination of such biochemical data as blood-sugar, respiratory quotient, etc. Having done that service to the enterprise, MacLeod left Canada for the summer, expecting, presumably, that workers with so little previous experience would have made, at most, some preliminary progress with their problem by the time of his return from Europe in the early autumn. He found, in fact, on his return, that Banting and Best had already demonstrated the possibility of solving a problem which had baffled so many investigators of ripe experience. MacLeod, as was natural, took the lead in providing for the rapid development of the *discovery* to practical fruition, by importing the collaboration of others. In all the circumstances, it was probably natural that the names of Banting and MacLeod should have been proposed to the Nobel Committee of those days for the award of the prize in 1923, without the mentioning of Best; and, if that indeed is what happened, the committee undoubtedly made the right and only possible decision in sharing the award between the two who were proposed. With some intimate knowledge of those concerned, however, dating from September 1922, when I visited Toronto (two months before Krogh) to acquaint myself on the spot with the new discovery I have no hesitation in offering the opinion that, if right proposals had been made, the inclusion of Best in the award would have been more in accordance with full scientific justice. Everything which I learned at that time convinced me that Best had at least an equal share with Banting in the essential discovery; and everything which has happened since has strengthened that conviction. MacLeod's share, so far as I was able to judge, had been to accelerate development by organising a team; nothing, I believe, was contributed thereby, which Banting and Best were not on the way to discovering for themselves, though they would probably have taken longer to do it. Since those earliest days, such further advances, towards the discovery of the nature of insulin and the mode of its action, as have been contributed by any of those who were concerned in any way with the original discovery, have been made by Best, and not by either Banting or MacLeod — both of whom of course are now dead. In further achievement also, over a much wider range of discovery, it seems to me that Best has long ago shown himself to be an investigator of higher rank than either Banting or MacLeod. I am venturing to suggest, therefore, to the present committee, that they might properly have in mind the doubt, whether full justice was done, to all concerned in the discovery of insulin, by the award in 1923, when they are considering Best's claim to an award, in this year, for a discovery which, though essentially different, actually began with observations which were first made in the course of the discovery of insulin and its action. I should add that, to my knowledge, Best himself remained, with conspicuous dignity, aloof from personal rivalries and contentions, which unfortunately arose among others concerned with the insulin discovery and its public recognition.

When reading this comprehensive nomination two things need to be considered. One is that Best and Dale were very close friends[29]. Best did his doctorate work under Dale, who even after that continued to be a solid supporter of Best's work. The other thing is that one can assume that Dale for tactical reasons over-emphasized Best's role in the discovery of insulin in order to improve his chances of getting a prize for his choline work.

As mentioned, von Euler provided comments to Dale's remarks on the insulin discovery in an appendix to his full evaluations in 1950 of Best's contributions after 1923. In this context it might be added that von Euler came to play a major role in the development of the Nobel Prize institutions. He was an ordinary member of the Nobel Committee at the Karolinska Institute 1955–1960 and its chairman for the last three years. In 1963–1964 he was an adjunct member, after which he became secretary of the committee. He then became distanced from the work, presumably because his own eligibility for a prize became apparent. Instead he served as the Chairman of the Nobel Foundation between 1965 and 1975 and it was during this time, in 1970, that he was awarded a prize in Physiology or Medicine, together with Bernard Katz and Julius Axelrod, "for their discoveries concerning the humoral transmittors in the nerve terminals and the mechanism for their storage, release and inactivation." It must have been a special occasion for him to be both the host of the events and one of the guests of honor during that year.

In his one and a half page appendix von Euler revisited the early stages of the discovery of insulin. He noted again that the original idea for the experiments came from Banting, but that Best, who had worked for a year on diabetes, made important contributions to the development of chemical procedures to extract insulin from the atrophied glandular tissue. He emphasized that the four original publications were co-authored only by Banting and Best. MacLeod's name was added to one of them at the time of the original presentation at the American Physiological Society meeting in New Haven in December 1921. The reason was that he was the only one who was a member of the society, a requirement for giving a presentation. It is von Euler's conclusion that the discovery was made by Banting and Best and that MacLeod did not make any critical contribution to its conceptualization. *Sic transit gloria mundi.*

Best continued to cultivate his contacts with the Swedish medical community. According to Bliss[29] he developed what is sometimes referred to as *Nobelomania*, a not uncommon syndrome of prestige-seeking scientists. He was a guest of honor at the Karolinska Institute in 1960 in connection with its 150 years' jubilee and in 1961 he was elected foreign member of the class of medicine of the Royal Swedish Academy of Sciences. His election was recommended by a number of Swedish members with a major influence on the Nobel work at the Institute, like von Euler and Liljestrand.

I am convinced that — had Banting's contribution been evaluated at a later time — the conclusion of the committee would have been that he alone should receive the prize. I doubt that the 24-year-old Best would have been included. He did make critical contributions in the development of a useful experimental insulin preparation, but the project was launched because of Banting's original idea and his drive to pursue it experimentally. Together with different collaborators he developed the experiments all the way to the testing in patients, where spectacular results were obtained. The young Best was important, but, as described by Bliss[25], Banting on occasions had to bring him back to order when he got too distracted by the other temptations of a young man. The fact that Best developed into a highly successful scientist and made discoveries considered prize-worthy is a separate issue.

A Nobel Prize awarded in the midst of World War II

The importance of a secretary with insight

In the midst of World War II the 1943 Nobel Prize in Physiology or Medicine was awarded in 1944 to Henrik C. P. Dam and Edward A. Doisy for their studies of vitamin K. This is the only prize given in a year when both candidates, in the absence of external nominations, were submitted by the secretary of the committee. Clearly the international turmoil was very influential. There was a precipitous drop in nominations and the number of candidates proposed during World War II (Figure 2). Dam had previously received one nomination in 1941, two in 1942 and two in 1943, Doisy had a single external nomination back in 1941. It was J. P. Strömbeck, professor of surgery at Lund University, who proposed Dam in 1941. He mentioned that

> After Dam's discovery a number of scientists have been involved in elucidating the chemical constitution of vitamin K and have produced other compounds with an effect similar to that of vitamin K and with great therapeutic value. In this context can be mentioned E. A. Doisy, L. F. Fieser and H. J. Almquist. I therefore want to raise the question if one or more of these would deserve to share the prize with Dam.

The committee took this as a nomination of the three candidates. Obviously Dam was the central figure in this discovery and the question to be answered was whether any of the chemists involved in the competitive race in 1939 to elucidate the chemical structure of vitamin K should be included. All four candidates were subjected to a full investigation by Eric Jorpes, professor of medical chemistry, and Adolf Lichtenstein, professor of pediatrics. Jorpes, in an exemplary and thorough evaluation, reached the conclusion that Dam was a strong and obvious candidate for a prize. He also recommended that Doisy should be included, since he had made the most pioneering and original contributions to insights into the chemical nature of the vitamin. Almquist and Fieser were not judged worthy of a prize.

Lichtenstein's conclusion differs, in part. He agreed that Dam was an outstanding candidate but suggested that Almquist might be included. He stated "without any hesitation" that Almquist should be placed ahead of the other researchers competing to join Dam in a prize. The committee's unanimous conclusion was that Dam was worthy of a prize and a majority (three reservations) found that this was also the case for Doisy. Almquist and Fieser were not thought to deserve a prize.

In 1942 and 1943 only Dam was nominated, and, supported by recommendations by two new reviewers, he was judged to be deserving of a prize. The committee then unanimously recommended that the 1943 Nobel Prize in Physiology or Medicine be given to Dam, but the faculty decided that no prize should be given this year, something that actually turned out in Doisy's favor. In 1944 there were no external nominations for discoveries concerning vitamin K. Liljestrand then nominated Dam and Doisy, following the 1941 recommendation by Jorpes. Another two full investigations of both Dam and Doisy were made by Hammarsten, and Arvid Wallgren, another professor of pediatrics.

Hammarsten agreed with the previous chemist evaluator Jorpes and endorsed the prize being given to both Dam and Doisy. Wallgren disagreed. He concluded that Doisy's contributions did not equal those of Dam and the latter alone should receive the prize. The committee then decided in 1944 that the 1943 prize should go to both Dam and Doisy. The faculty agreed and gave a divided prize to Dam and Doisy, the reasons being, respectively, "for his discovery of vitamin K," and "for his discovery of the chemical nature of vitamin K."

When Almquist learned that Dam had received the Nobel Prize he sent him a letter of congratulation. Dam, being a very modest and unpretentious person[30], responded to this and noted, "There must have been reasons for some bitterness for you in the fact that you so nearly missed being the first to report the existence of vitamin K." Almquist naturally did feel "some bitterness" and argued that it was he, and not Doisy, who should have shared the prize with Dam.

Prize award in the diaspora

On December 10, 1944, six of the eight individuals awarded the scientific or literary prizes, including Dam, were in the US as citizens or refugees. Dam

Henrik Dam (1895–1976) and Edward Doisy (1893–1986), recipients of the Nobel Prize in Physiology or Medicine 1944. [From *Les Prix Nobel 1940–1944*.]

had fled soon after April 9, 1940, when Denmark was occupied by German troops. He spent most of his time at Rochester, New York, and did not return to Denmark until June 1946. Dam remained a private and rather reserved person. In 1946 he was installed in a chair at the Technical University in Copenhagen, to which he had already been appointed in 1941. He continued to do high quality work but remained relatively unknown outside the small community of Danish biochemists and nutritional experts.

There was no Nobel Prize ceremony in 1944 in Stockholm, but a celebration was arranged in New York under the auspices of the American-Scandinavian Foundation, and prizes were handed over by the Swedish Minister in Washington. In addition the prize recipients were greeted via Swedish radio by His Royal Highness, Crown Prince Gustaf Adolf, and by the Chairman of the Nobel Foundation, The Svedberg. By the same means Lichtenstein gave a laudatory speech for Dam and Doisy.

The prize for their discovery of vitamin K was the last in a series of prizes awarded to nine persons between 1928 and 1943 for studies of vitamins[30], a term introduced in 1912 by the Polish chemist Casimir Funk. It was originally believed, incorrectly as was later demonstrated, that they all would be amines, thus containing nitrogen. Several important discoveries in vitamin research were never recognized by a prize, like that of vitamin D. In 1928 it was concluded that the discovery of this vitamin was a great advance, but the credit had to be shared among so many scientists that it was considered an impossible task to select a few individuals for an award.

A candidate who almost did not make it

The discovery of penicillin is one of the most obvious findings to be awarded a Nobel Prize in Physiology or Medicine. Its serendipitous discovery is described in Chapter 2. As material for use in patients became available during World War II and its remarkable effect was quickly demonstrated it was inevitable that it should, sooner or later, be recognized. In 1943 one nomination arrived for Alexander Fleming and Howard A. Florey and in 1944 one for Fleming alone and another for Fleming together with Ernst B. Chain and Florey. In 1945 there was an avalanche of 18 nominations followed with an additional 10 nominations submitted after the deadline of January 31. All the 18 nominations were for Fleming alone or for Fleming in combination with Florey. The only nomination including Chain was by the secretary of the committee.

A preliminary examination of Fleming and Florey was made in 1943 by Liljestrand. He praised Fleming's discovery and stated that it might be a major advance in the field of chemotherapeutics. However, he believed that more confirmatory data needed to be accumulated and did not recommend a full investigation. This was also the committee's conclusion. In 1944 Fleming and Florey were evaluated again, now together with Chain, who was included in a single nomination by Professor E. W. Whitehead from Denver, Colorado. It should be remembered that in 1944 the nominations for Nobel Prizes in Physiology or Medicine were at a *nadir* (Figure 2). The committee decided that a full evaluation should be made by two reviewers, Liljestrand again, and Nanna Svartz, professor of medicine.

Fleming's classical original discovery is carefully described. Following up his original finding in 1922 of lysozyme as an effective antibacterial ferment, he observed in 1929 that a certain mold would interfere with the growth of Staphylococci. In order to observe bacterial colonies with a magnifying glass one had to remove the lid from the Petri dish where they were growing, and this had allowed for accidental fungal contamination of the culture from the laboratory environment. Instead of throwing the plate away Fleming became curious and extended the studies of the antibacterial substance secreted by the unique mold (see also Chapter 2). It turned out to be a relatively rare species previously classified by a Swede, R. P. Westling. It had been named *Penicillin notatum* and Fleming therefore named the substance penicillin. Although the antibacterial effect had been published by Fleming already in 1932 it took until World War II before penicillin could be prepared in useful quantities, be chemically characterized and finally synthesized. Liljestrand concluded that it was Fleming who had made the primary discovery and that he, without doubt, was worthy of a prize. He also came to the conclusion that Chain and Florey had done the work that eventually led to the production of penicillin of a quantity and quality that allowed it to be used therapeutically. Both were judged to deserve a prize, a conclusion shared by Svartz.

This was confirmed in part by the committee, which, however, was divided in its recommendation for the 1944 prize. Four members voted for giving it with one half in favor of Fleming and the other half in favor of Chain and Florey, but the remaining four instead recommended Joseph Erlanger and Herbert S. Gasser. The faculty decided to give the prize to the latter candidates "for their discoveries relating to the highly differentiated functions of single nerve fibres."

The flooding of the committee with nominations for Fleming and Florey in 1945 elicited three more investigations. These included Chain, since Liljestrand had nominated all three candidates, referring to the investigations of 1944. The evaluations were made by Sven Hellerström, professor of dermatology and venereology, Anders Kristensson, and Liljestrand. The three endorsed all the candidates, but recommended that Fleming receive half the prize — since he had made the original discovery — and Chain and Florey the other half. This proposal was submitted by the committee to the faculty. Eventually, however, the faculty decided not to split the prize in two, but to award the prize in equal shares jointly to Fleming, Chain and Florey (listed in this order), "for the discovery of penicillin and its curative effect in various diseases." Liljestrand gave the laudatory address at the prize ceremony[31] (one out of the seven that he gave during his 42 years as secretary) and there were reasons for Chain to be particularly grateful to him. He became the critical arbiter in both the 1943(4) and 1945 World War II prizes.

It seems that from a historical perspective the equal sharing of the 1945 prize among the three laureates was entirely justified. There is no question about the fact that Fleming did the original observation, but it is difficult to understand why he did not pursue this observation with more vigor between 1929 and 1941. In the early 1930s a group of natural-product chemists worked on penicillin, but they gave up due to the instability of the compound. It was in 1939 that scientists at the Sir William Dunn School of Pathology at Oxford decided to pursue the development of antibacterial compounds and after a few years were successful in producing penicillin of sufficient quality and quantity to be tried in patients. Fleming's original observation became

From left: Alexander Fleming (1881–1955), Ernst Chain (1906–1979), and Howard Florey (1898–1968), recipients of the Nobel Prize in Physiology or Medicine 1945. [From *Les Prix Nobel 1945.*]

possible to resurrect. The initiative to undertake this work was a joint effort by Chain and Florey, the head of the institution. It was the combination of Florey's biological knowledge and Chain's competence as a biochemist that allowed the determination of the chemical nature of the antibacterial compound and the successful full-scale production of purified penicillin in larger quantities.

On August 5, 1942 Fleming returned to his engagement in the possibility of the treatment of infections with penicillin. In a telephone contact with Florey he inquired about the possibility of getting access to the first lot of purified material for treatment of a personal friend who was dying of streptococcal meningitis at St. Mary's Hospital where Fleming was active. Florey responded positively and personally took the entire stock of penicillin available at the moment by train to London. The patient miraculously recovered after an intraspinal injection of the compound. This was the origin of the "Fleming myth," which caused the media for a long time to overlook the absolutely indispensable contributions by Florey and Chain[32]. The panegyric praising of Fleming in the press at the time resulted in very strained relationships between the two institutions in London and Oxford. It is praise-worthy that the Nobel Committee at the time, in its recommendation for prize recipients, could rise above this kind of conflict. They are, regrettably, not uncommon in the academic field in situations concerning allocation of priorities of a discovery and of giving fair credit to all scientists involved.

DDT, a uniquely effective insecticide which fell into disrepute

DDT marches with the troops

During the early decades of Nobel Prizes in Physiology or Medicine, combating infectious diseases was often the center of interest[2,5,6]. The need to find effective means to control them became particularly apparent during wartime conditions. In 1928 Charles J. H. Nicolle was awarded a prize "for his *work on* (not discovery of) typhus." He showed the importance of the body louse in the transmission of the disease. The importance of hygiene was fully appreciated, but there remained a need to find chemicals that could eliminate the vector of the agent. Towards the end of World Was II it was shown that DDT (Dichloro-Diphenyl-Trichloroethane) offered unique possibilities for vector control. It became critical to the operations of the Allied army and in fact saved many hundreds of thousands of lives. In 1942, 100 kilograms of Geigy's new insecticide Gesarol, which contained DDT, was sent from Switzerland to the US, and its impressive effects were rapidly confirmed in the Orlando laboratory of the US Department of Agriculture and the Louse Laboratory of the Rockefeller Foundation in New York. The latter laboratory conducted field trials with a group of conscientious objectors assigned to a forestry camp in New Hampshire, which they had been put into contact with through a Quaker agency[33,34]. From then on it can be said, "DDT marches with the troops."

The committee and faculty moved quickly in 1948

In October 1947 the Nobel Committee for Physiology or Medicine, following its routine rotating scheme for asking universities around the world for proposals, invited the medical faculty of Istanbul to give nominations for the 1948 prize. Seven nominations were received for Paul H. Müller at J. R. Geigy Dye-Factory Co., for his discovery of DDT. Their form differed with regard to both their length and the language used. However their basic structure was the same and it is obvious that they represented a single, coordinated proposal.

The committee selected Gunnar A. V. Fischer, professor of public health, to review the proposals. In his report, he first noted that, two years earlier, the honorary doctor Paul Läuger from the same company had been nominated for the discovery of DDT. This nomination had been subjected to a full analysis by himself and by Liljestrand. Both had concluded that the discovery of DDT was worthy of a prize, but that it had been made by Müller and not by Läuger, and that the committee had to wait for a correct nomination. Läuger was the head of a unit at the research laboratory of the J. R. Geigy Dye-Factory Co., where interesting compounds to counter textile damage by insects had been developed during the 1930s. The development of DDT was seen as a separate and independent initiative by Müller, a co-worker in the laboratory. Thus, in contrast to the situation in 1923, for the discovery of insulin — it was deemed important to directly award the scientist who had made the discovery, and not the head of the laboratory.

In systematic studies of derivates of diphenyltrichloroethane, Müller had synthesized DDT in 1939. Actually, the compound had been synthesized already in 1873 by an Austrian student, Othmar Zeidler, as a part of his doctorate thesis. The novelty was that it was found to be a highly efficient insecticide. By the criteria Müller applied, it seemed close to ideal; it killed a broad spectrum of insects, it was effective in low doses and was highly stable. Furthermore it was easy and inexpensive to manufacture. Because of its lipophilic properties, DDT is rapidly absorbed by insects and exerts its toxic effects as a contact poison. Geigy took out a patent on the compound in 1940.

In his assessment, Fischer discussed the field tests performed with DDT and the different sensitivities to the compound of various insect species. Mosquitoes, fleas and body lice were highly sensitive, whereas bed bugs and cockroaches were rather resistant. The common fly showed variable resistance and the emergence of resistance was observed, which was not found in mosquitoes and body lice. Toxicity for man appeared low, and no symptoms were observed when the compound was used at recommended concentrations. The effects of heavy exposure needed further studies. A number of examples were given on the powerful effects of DDT on eliminating the body louse, with ensuing dramatic consequences of typhus epidemics in various conditions of war, also including concentration camps. A multitude of studies of the effect on the spread of malaria after DDT use in many parts of the world was also discussed. Finally, Fischer described the effect on a number of other vector-borne diseases, such as yellow fever.

He summarized the discovery of DDT and its value for elimination of diseases as follows:

Thus DDT has been shown to be of great value in the work to prevent a large number of the arthropod-borne diseases, like typhus and malaria, which among the epidemic diseases in general cause the largest number of diseases and deaths. DDT has certainly already saved the health and lives of hundreds of thousands of humans, and for the future the new DDT technology opens unforeseeable aspects.

With reference to what has been said above I conclude that Paul Müller's discovery of the strong effect of DDT as a contact poison is worthy of a prize.

In 1948 the committee solicited full investigations of 12 research areas and among the reports received it singled out the discovery of DDT. The concluding statement was: "The Nobel Committee has decided unanimously to propose that the 1948 year prize in Physiology or Medicine be awarded to Paul Müller 'for his discovery of the strong effect of DDT as a contact poison against a number of different arthropods'." This also became the decision of the faculty.

Paul Müller (1899–1965), recipient of the Nobel Prize in Physiology or Medicine 1948. [From *Les Prix Nobel 1948.*]

In his laudatory speech at the prize ceremony, Fischer[35] recapitulated the discovery and the dramatic effects of the use of the compound. Its last paragraph reads:

The history of DDT illustrates the often very crooked way forward towards a major discovery. A researcher working with flies and Colorado-beetles finds a compound which turns out to be effective in the fight against the most difficult diseases in the world. Some may say that he has been lucky, and that is of course true. Without a good deal of luck it is barely possible to make major discoveries. But it is not only a question of luck. The critical observation has been made during the course of assiduous and certainly often monotonous work, and it is the capacity to see, appreciate and value the importance of what seems to be an insignificant discovery, which constitutes the true scientist.

Müller's Nobel lecture[36] describes the often undistinguished life of an industrial chemist, and emphasizes the role of a systematic, persistent and imaginative approach. He remained with the company where he had started his professional career in 1925. Before his death in 1965 he would witness some re-evaluations of the usefulness of DDT.

The future saga of DDT

The use of DDT was expanded after World War II, in particular for the control of malaria. An early extensive field testing was supervised by the Rockefeller Foundation in Sardinia; after saturating the island with DDT the disease had disappeared. The expanded use of the insecticide contributed to the eradication of malaria from Europe and North America. The encouraging results stimulated the World Health Organization (WHO) to become involved, and in 1955 it launched its program for worldwide malaria eradication. Encouraging results were obtained, with a marked reduction of mortality rates. But problems soon started to emerge and it became apparent that DDT was not a simple, single solution to the control of vector-borne diseases. Insects that had evolved resistance to DDT started to be recognized after the campaigns. Also the toxicity of the chemical had been underestimated, and its impact on the environment started to become visible.

Discussions began about regulating the use of DDT, and *The New York Times* editor William Shawn brought this problem to the attention of the popular naturalist-author Rachel Carson. Her engagement led to publication, in 1962, of the best-selling book *Silent Spring*[37] — one of the most influential books written in the 20th century, the publication of which launched the modern environmental movement. Carson was never to see its full impact; she died of cancer in 1964.

Soon the potential harm from the use of DDT was being hotly discussed[38]. It was appreciated that the compound stayed in the organism for a long time, because of its lipophilic nature, and as a consequence it accumulated in the food chain. Birds of prey, in particular fish eaters, became seriously affected. There were also arguments that the toxicity for man had been underrated, but this was poorly documented. The possible (marginal) carcinogenic effect of DDT is still a matter of debate. The goal of the WHO to eradicate malaria was abandoned in 1969. The agricultural use of DDT was banned first in Norway and Sweden in 1970, in the US in 1972 but in the UK not until 1984. The Stockholm Convention, ratified in 2001 and effective as of 17 May 2004, calls for the elimination of DDT and other persistent organic pollutants.

However, solutions to life's problems are rarely black and white. In particular DDT is still used today and the WHO continues to recognize it as one important tool in the fight against malaria. The difference in its use is that it is now applied inside buildings and in other settings only by selective spraying, all with the intention of reducing the environmental impact. DDT is also used for impregnation of nets used to provide protection against mosquitoes carrying malaria parasites. In the end it is a question of balancing the benefits and the risks, and clearly many lives can be saved by a considerate use of DDT. There is hope for an even more efficient and harmless insecticide, but until such is available, the DDT compound that was awarded the 1948 Nobel Prize in Physiology or Medicine will continue to be used — though only selectively.

Corticosteroids, another medical wonder drug

In 1950 the Nobel Prize in Physiology or Medicine was given to Kendall, Tadeus Reichstein, and Philip S. Hench, for discoveries which had been published during the preceding year. It had been found that substances from the adrenal gland cortex and also pituitary gland extract that stimulate the cortex, Adreno-Cortico-Tropic-Hormone (ACTH), had a spectacular curative effect in patients with rheumatoid arthritis[39].

Kendall was well known to the committee, having been nominated for the first time in 1922 for the purification of thyroxine, the hormone of the thyroid gland, and description of its chemical structure, a proposal repeated seven times until 1938. He was subjected to four full investigations and his contribution was judged to be worthy of a prize. However, he never made it to the top of the list. In 1941 there were four more nominations for Kendall, but now for work on the chemical composition and physiological action of the hormones of the adrenal cortex. Hugo Theorell, professor of biochemistry, and forthcoming 1955 Nobel Laureate in Physiology or Medicine, performed the evaluation. Theorell reviewed Kendall's steroid research at large and emphasized that his contributions should be compared with those of Tadeus Reichstein and Oskar Wintersteiner, two scientists not nominated for the prize. His conclusion was that the chemical studies of Reichstein matched those made by Kendall, but that the latter had contributed more to the understanding of the physiological functions of the compounds. Theorell states, with some hesitation, that Kendall would be deserving of a prize, if his earlier studies of thyroxin were included. The committee did not follow this suggestion.

As in the case of penicillin in 1945, the discovery, in 1949, of dramatic clinical effects of corticosteroids and ACTH on rheumatoid arthritis, prompted a new wave of nominations of Kendall in 1950. He was proposed 14 times and eight of the nominations also included Hench, for the first time this year. In addition Hench alone got three nominations. The third candidate, the chemist Reichstein, was also proposed for the first time in 1950, by two nominators: J. M. Yoffey, professor of anatomy in Bristol (together with Kendall), and by a member of the committee, Jorpes (together again with Kendall and in addition Wintersteiner, the other important chemist engaged in purifying corticosteroids). Yoffey was hesitant in his nomination and stated that "I do not feel really competent either to arrange the various workers in order of merit or even to present in detail the work of … ." Jorpes' nomination had a somewhat different focus than the others, namely for "the discovery of the nature of adrenal cortex hormones and clarification of their chemistry." The committee decided to let Hammarsten, who had evaluated Kendall way back around 1930, write a new full review. He also provided a four-page report on the first-time nominee Reichstein. Further in-depth evaluations were made of Kendall, now together with Hench, by Liljestrand and Svartz.

The pronounced clinical effect of steroids on rheumatoid arthritis was a discovery of the same rank as the discoveries and clinical applications of insulin and penicillin. It was therefore natural that the committee was tempted, in the spirit of Alfred Nobel's will, to quickly award this fresh discovery (published in April–May of 1949) in 1950, in spite of the fact that both Hench and Reichstein were nominated for the first time that year. Although a newcomer as a Nobel Prize candidate, Hench, as a leading clinician at the Mayo Clinic, Rochester, Minnesota, the same place where Kendall was working, was a legendary figure in rheumatoid arthritis research. He had made the important early clinical observation that under certain conditions, in particular a concomitant occurrence of pregnancy and jaundice, the symptoms of arthritis could improve, sometimes markedly. He became convinced that compounds with a capacity to counteract the disease potentially could be produced in the body.

Kendall's pioneering steroid research, in particular on what originally was called compound E, eventually led to it becoming available in sufficient amounts for a clinical trial in 1947. Chemists at Merck and Co., in particular Dr Sarett, were crucial in making this possible. The three evaluations were in full agreement and pronounced Kendall and Hench worthy of a prize, but

From left: Edward Kendall (1886–1972), Tadeus Reichstein (1897–1996), and Philip Hench (1896–1965), recipients of the Nobel Prize in Physiology or Medicine 1950. [From *Les Prix Nobel 1950.*]

Liljestrand was hesitant and noted, "it is still too early to decide if cortisone and ACTH will have practical importance," and further, "According to my opinion the therapeutic importance of the contribution of the Mayo Clinic is not such … that it motivates a Nobel Prize."

It seems that the final inclusion of Reichstein in the prize was somewhat shoehorned. Based on only two (weak?) nominations, he was subjected to a brief review by Hammarsten, whose conclusion reads: "I therefore consider that he has such a large part in Kendall's discovery of active steroids, especially substance E, that his contributions are worthy of a prize. However it is my opinion that Kendall's entitlement to a prize is stronger than Reichstein's."

It was not surprising that the committee had difficulties developing a unanimous proposal to present to the faculty. Nine of the 13 members supported a joint prize to Kendall, Reichstein and Hench, but four members instead wanted to give a divided prize to Max Theiler (laureate-to-be in 1951, Chapter 4) and F. Macfarlane Burnet (one of the two laureates-to-be in 1960). The faculty supported the majority of the committee and gave a prize, equally divided among the three candidates Kendall, Reichstein and Hench, "for their discoveries relating to the hormones of the adrenal cortex, their structure and biological effects." Thus the recent observation of a favorable effect on rheumatoid diseases was not mentioned. In his laudatory presentation at the prize ceremony[40], Liljestrand spent most of his time highlighting the development of the chemical knowledge about the different steroids, frequently mentioning both Kendall and Reichstein, and only briefly, towards the end, the recent clinical findings by Hench and his collaborators.

It is probably fortunate that the committee and the faculty did not wait for more clinical results. It did in fact take more than 10 years before a large clinical study could confirm that the effect of cortisone was superior to that of aspirin. Today, more than 50 years later, the treatment of rheumatoid diseases with steroids has stood the test of time[39].

Whereas the recognition of the importance of discoveries of vitamins, the critical factors depending on exogenous supply, came to an end with the award of the 1943 (1944) prize to Dam and Doisy, this was not the case for hormones by the award of the 1950 prize for corticosteroids. Due to particular circumstances[5] (previous chapter), a prize in Chemistry was given in 1955 to Vincent du Vigneaud[41] "for his work on biochemically important sulphur compounds, especially for the first synthesis of a polypeptide hormone." As late as 1977, a prize in Physiology or Medicine was given with one half to Roger Guillemin and Andrew V. Schally "for their discoveries concerning the peptide hormone production of the brain," and the other half to Rosalyn Yalow "for the development of radioimmunoassays of peptide hormones."

Joshua Lederberg, a rapidly recognized scientist

An inspiring nomination

One half of the 1958 Nobel Prize in Physiology or Medicine went to George W. Beadle and Edward L. Tatum "for their discovery that genes act by regulating definite chemical events" (next chapter). In contrast to these recipients, nominated since 1948, Joshua Lederberg, who received the other half of the prize, was proposed for the first time that same year. At the time of his recognition he was only 33 years old, the second youngest prize recipient in Physiology or Medicine ever (page 31).

Lederberg was nominated together with four other scientists by the Dean of the Faculty of Medicine at Harvard University, the professor of bacteriology George P. Berry. The five-page nomination gives a broad review of recent developments in the field of genetics. It notes, as a departure, that "No area of biology has had a more striking efflorescence in the past 20 years than has microbial genetics." Microorganisms provided ideal tools for genetic studies, and a host of brilliant scientists had become involved in the development of this new field. After 1958 many prizes in Physiology or Medicine and also Chemistry were awarded for the impressive discoveries in the field known as molecular biology, as discussed in the next chapter. Berry lists five advances recognizable in 1958. These are:

1. The late Oswald T. Avery's identification of DNA as the carrier of genetic information.
2. The use of microbial mutants to correlate gene and product functions and to construct genome maps shown by Beadle and Tatum.
3. The discovery of many kinds of exchange of genetic material between bacteria. Originally it was assumed that no such changes occur and that bacteria simply multiply by binary fission. This turned out to be wrong and bacteria became ideal systems to study mutation and exchange of genetic material (recombination). Another useful finding was that bacteriophages can transmit not only their own genomes, but also host genome fragments from one cell to the other (transduction).
4. One consequence of these breakthroughs was that genes could now be studied as functional units. It was postulated by Berry, correctly as it turns out, that this understanding would pave the way for gaining new insights into the molecular mechanisms governing inheritance.
5. Bacteriophages can infect and, as a consequence of their replication, kill bacteria. However, under special conditions, the viral genome may instead integrate into that of the cell and remain dormant until it becomes activated. This is referred to as a *lysogenic state* and the virus genome as the *prophage*. The demonstration of this phenomenon and clarification of the underlying mechanisms led to a conceptual softening of the distinction between a gene and a virus (Chapter 3).

The brilliant review of Lederberg's scientific contributions

The committee was impressed by the nomination of Lederberg and decided to let Georg Klein (Chapter 3), professor of tumor biology since the year before and a member of the committee for the first time, make a full assessment. Klein, who was born in the same year as Lederberg, in 1925, wrote a 25-page review (plus a reference list of more than 70 articles), which is splendid both in content and form. It can serve as another prototype of high-class Nobel Prize candidate reviews.

Lederberg was a very successful young scientist who, in 1946–1947, had had Tatum as his critical mentor at Yale University, where he also received his

PhD. He quickly became an independent group leader at Madison/Wisconsin University, pioneering a number of discoveries. Klein lucidly reviews Lederberg's contributions and carefully allocates priorities to the different findings. He then concluded that Lederberg was qualified for a Nobel Prize by his discoveries both of the phenomena of genetic recombination, transduction and lysogeny. Klein argued that the most natural combination would be to combine Lederberg with André Lwoff, who had also made critical observations leading to an understanding of lysogeny, and had been investigated by Sven Gard and declared worthy of a prize in 1955[42] (Chapter 3).

Klein had related to me how he as a young professor and a newcomer as adjunct member of the Nobel Committee in 1958, he listened to the intense arguments between chemists and physiologists for prioritizing of prize candidates. The young Sune Bergström (Chapter 1) sitting at his side proposed discreetly, "Why don't you argue for the geneticists?" Apparently Klein's intervention was well taken because in the end the committee recommended to the faculty that half of the 1958 prize should be given to Beadle and Tatum and the other half to Lederberg "for his discoveries concerning genetic recombination and the organization of the genetic material of bacteria." There was only one dissenting voice, viz. Gard, who preferred the combination of Lederberg and Lwoff. The faculty agreed with the overwhelming majority of the committee and Lwoff had to wait until 1965 when he shared a prize with François Jacob and Jacques Monod.

In his elegant Nobel lecture[43] entitled *A view of genetics,* delivered six months after the prize ceremony, Lederberg took the opportunity not only to review the progress made in his research but also to make general visionary statements. The lecture includes sections on DNA as the central gene-carrying structure, on viruses versus genes (Chapter 3) and on the creation of life. It is obvious from his lecture that eventually DNA had come of age as the genetic material (Chapter 7).

Joshua Lederberg (1925– 2008), recipient of one half of the Nobel Prize in Physiology or Medicine 1958. [From *Les Prix Nobel 1958.*]

In the year in which he received his Nobel Prize, Lederberg moved to Stanford University, where he became the founder and chairman of the Department of Genetics. Besides pursuing his research he had many engagements and frequently served as an advisor to the US government. He also became involved in the field of exobiology and was concerned about the risks associated with the potential accidental transport of microorganisms to and from Earth through space flights. Another interest of his was artificial intelligence. From 1978 to1990 he served as President of the Rockefeller University. He died in 2008.

The formative years in the work of the committee

This chapter discusses a selection of ten awards from among the first 50 prizes in Physiology or Medicine. These ten awards and the three separately discussed awards[3,4] (Theiler, Weller-Robbins, Kornberg — Chapters 4, 5, 7) concern prizes for which the Nobel Committee had taken particular initiatives. In most of these cases the selection procedure was accelerated to allow the acknowledgement of recently made discoveries or cases considered particularly urgent to recognize. Events connected with these prizes highlight a number of the central issues to be considered by the Nobel committees and the faculty at the Karolinska Institute. There were many things to reflect on when it came to the specifications given in Nobel's will. One was to interpret what should be included within "physiology or medicine."

The two terms must be understood in the context of the improvement of human health. The term physiology on its own is a very wide concept and encompasses the science which treats the functions of the living organism (of *any* kind) and its parts, and of the physical and chemical factors and processes involved. In general the Karolinska Institute, wisely, has taken a very liberal view on the interpretation of "physiology or medicine" as emphasized by Liljestrand[2]. It has simply been taken to mean all the theoretical as well as practical medical sciences, encompassed by the more modern terms biomedical or life sciences. With time it has become appreciated that different forms of living organisms are more closely related than originally believed. Recent insights into the structure of genomes of different species highlight this fact. Therefore information of value for the improvement of human health can sometimes be gained from studies of relatively simple systems. Genes of importance for the development of cancers, or exceptional infectious agents like prions, can be studied in yeast cells (Chapter 8). We live in a world of bacteria and viruses, some of which may cause diseases, while others serve various beneficial functions. Hence we need to know how they function and interact with our body and our immune system[42] (Chapter 3).

The terms "physiology or medicine" in the prize definition overlap to some extent, because the experimental science of physiology often goes hand in hand with the clinical evaluations of new interventions to further human health. Not infrequently has the "or" been exchanged for an "and." There is still another overlap, namely between the fields covered by prizes in Chemistry and in Physiology or Medicine. Originally it was up to each prize-awarding institution to decide which prize within the domains of organic chemistry and life sciences one would select. Thus the first prize in virology was half a Chemistry prize in 1946 to John H. Northrop and Wendell M. Stanley and a second prize in the same field for studies of tobacco mosaic virus was given to Aaron Klug in 1982. All other prizes for studies of virus systems were given in Physiology or Medicine[42] (Chapter 3). Another example is the choice between Theorell and du Vigneau for prizes in Chemistry and Physiology or Medicine in 1955 discussed in Chapter 5.

In the field of molecular biology the prizes in the beginning were given in Physiology or Medicine, but later in both this field and in Chemistry as will be further discussed in the next chapter. For many decades past the Nobel Committees for Physiology or Medicine of the Karolinska Institute and the Nobel Committees for Chemistry of the Royal Swedish Academy of Sciences have had joint meetings to decide about how to distribute responsibilities for candidates from different subdisciplines of science.

Still in 1959 the committee at the Karolinska Institute allocated the nominations to one of six subdisciplines within physiology or medicine. Such a subdivision of candidates has long since been abandoned, but it is still used by the committees for Physics and for Chemistry at the Royal Academy of Sciences. There is no rotation between subdisciplines when it comes to the selection of candidates. It is simply the unique merits of a candidate that count in the final competition. Since there tend to be several very qualified candidates, it is less likely that the same subfield is recognized two years in a row.

Many recipients of the prize in Physiology or Medicine have had no training in medicine, nor have they been members of medical faculties. This is not surprising, since the terms physiology or medicine have been broadly interpreted. Several of the laureates discussed in this chapter had a non-medical background. In addition to Krogh, who belonged to this category, there were chemists like Hill, Dam, Doisy, Chain, Florey, Müller, Kendall and Reichstein. To this list can be added Albert von Szent-Györgyi Nagyrapolt, Selman A. Waksman and Daniel Bovet, who received their prizes in 1937, 1952 and 1957, respectively. Prizes to non-medical scientists were also given in the field of genetics and developmental biology. None of the 1958 prize recipients discussed above had an M.D. degree and prior to them Thomas H. Morgan, Hans Spemann and Herman J. Muller, prize recipients in 1933, 1935 and 1946, respectively, lacked training in medicine. Altogether 21 of the 76 scientists who were recognized by a prize in Physiology or Medicine between 1901 and 1959 did not have a medical background.

The specification "conferred the greatest benefit to mankind" was also a subject of reflection in the early phase of prize selection. To what extent should prizes recognize applied sciences? It was soon accepted that prizes could — and should — be given to very basic discoveries as well. When Johansson became chairman of the committee, he emphasized that in the selection process, major consideration should be given to originality and to the importance in a scientific context[1]. This attitude has been progressively strengthened with time and most prizes are given for basic discoveries, sometimes to the disappointment of members in the Nobel Assembly with a clinical background.

The formulation "during the preceding year" was the source of more extensive discussions. In three particular cases reviewed in this chapter, Krogh, Banting and MacLeod and Kendall, Reichstein and Hench, the faculty managed to award a discovery that had been published the preceding year, but this has not been so in the other cases. A certain delay in awarding a prize is generally wise. The committee needs to fully understand the magnitude and robustness of the discovery and properly identify the priorities claimed by the scientists involved in the new findings. The latter, understandably, is not an easy task, but with time the committees have developed highly efficient procedures for impartial comprehensive, high-quality evaluations. This tradition has its origin in the way in which the selection process was designed as early as the first year in which a prize was awarded, in 1901. The delay in recognizing a discovery is sanctioned by a provision in the statutes that reads, "The provision in the will that the annual award of prizes shall refer to works *during the preceding year* shall be understood in the sense that the awards shall be made for the most recent achievements in the field of culture referred to in the will, and for older works only if their significance has not become apparent until recently." Admittedly, in some cases, as in the prize to Sherrington in 1932 and Rous in 1966 (Chapter 3), the committee and faculty at the Karolinska Institute did indeed stretch the meaning of "become apparent."

The most important and ingenious specification in the will is that the prize in Physiology or Medicine shall be given for a *discovery*. As pointed out this is narrower than in Physics and Chemistry which also allow awards for inventions and improvements, respectively. The meaning, content and delimitation of the term "discovery" have been the subject of many discussions of the Nobel Committee and Faculty at the Karolinska Institute. These discussions still go on today and will continue into the future. What constitutes a clearly defined discovery? How does it evolve by rational science and serendipity (luck)? When is a discovery complete, etc.? The questions are many and they need to be rigorously addressed. Some aspects of this issue were given in Chapter 4[3]. It is not uncommon that a committee rewards the combination of an original observation and the follow-up developments that lead to the availability of a useful compound or diagnostic test, as in the cases of vitamin B12, vitamin K, penicillin and corticosteroids. In the more recent 2008 prize, however, only the original isolation of human immunodeficiency virus (HIV) was recognized by the awarding of the prize to Françoise Barré-Sinoussi and Luc Montagnier, and not the important applications of this knowledge.

The Nobel Prizes are unique in that they got off to a start more than 100 years ago; that from the very beginning they were international and that they contained a substantial financial reward. The selection process obviously has evolved, but was of high calibre already from the beginning. Many excellent written proposals have been received from the selected group of invited nominators and they have been screened efficiently. It was only during World Wars I and II that the flow of nominations dried up. The strongest candidates were subjected to very qualified reviews. These reviews originally were made mostly by members of the committee, but very soon other members of the faculty became involved. Eventually national and to a certain extent international reviewers also became engaged.

Nobel committees are specifically required to have three to five members. Originally all of them had five members, but at the Karolinska Institute a lower number was used later. During the Second World War the committee was reduced to three members and not until 1961 did it again have five. The secretary is present either as a voting or a non-voting member at committee meetings, a situation which for some periods has also applied to the Vice-Chancellor of the Karolinska Institute. However, the field of biomedical research is wide and to manage this, the operating committee was expanded by the temporary inclusion of faculty members providing the required overview of diverse fields. Hence already in the 1930s the operating committee included some ten people. Since 1970 there has been a yearly adjunction for eight months (March through September) of ten people, selected to represent the fields with candidates judged to be of the highest relevance for the work of that year, as already mentioned.

This enlarged committee comprising a total of 15 members reports, during the later part of September, to a Nobel Assembly of 50 members. This Assembly took over responsibilities from the faculty in 1977 (Chapter 1).

A critical part of the evaluation of candidates is the integrity and secrecy of the process. This has developed and been maintained surprisingly well. It is emphasized already to the individuals invited to nominate candidates that they must keep the nomination secret. The persons involved in the selections are well aware of their responsibilities and the procedures are anchored in a long tradition of impartial, objective and non-partisan evaluations; they are ingrained in the cultural (Lutheran) traditions of two small and homogeneous, sparsely-populated countries on the outskirts of Europe. It may have been that Nobel showed a fortunate insight when he decided to let Sweden and Norway carry the responsibility for the prizes and not the larger European country where he lived for a good part of his life. A secret, immutable selection process, essentially not accessible to any external influence, has evolved. Naturally the choices of laureates are sometimes criticized, but the prize-awarding institutions never respond to this.

The archive materials of the Nobel Committee at the Karolinska Institute, which may be released for scientific historical evaluations when the 50 years' secrecy period has expired, provide excellent illustrations of the meticulous work and deliberations of the committees. The protocols of the meetings are brief since it is not allowed to take notes of the different opinions raised. Clearly one often finds divergence of opinions or can sense the competition for the first place of different disciplines, but in the end a unanimous or at least a solid majority proposal is generally made to the faculty. The faculty does not always follow the recommendation of the committee and sometimes requests for reconsideration have been sent back, as illustrated by some cases discussed in this chapter. The outcome of the selection process has generally been well accepted by the international scientific community and quite early the prize was taken as the ultimate accolade among international scientific awards. The discoveries awarded truly have highlighted the milestones of modern biomedical research. The prestige of the prize has continued to increase with time, which in turn put additional pressure on those involved in the selection process. There is simply no room for mistakes. The amount of high-quality science published today and the number of scientists involved are orders of magnitude larger than a hundred years ago, which adds to the challenges of the selection task.

No more than three scientists can share a prize, which sometimes poses difficulties for the committee. In the beginning mostly individual recipients were awarded. There were a few prizes bestowed upon two recipients, but it took until 1934 for three recipients to receive a prize (Table 1, Chapter 1). One important rule of thumb when it comes to sharing is to evaluate if each of the candidates proposed to be included in a prize would be qualified to carry a prize on his own. This principle has not always been adhered to. There are examples, as in the case of the discoveries of vitamin K, penicillin and steroids, when one candidate has been considered much stronger than another, although in the end the award was shared equally.

This chapter focuses on cases when the committee has taken particular initiatives, where it pushed a candidate or has secured a nomination by one of its members, generally by the chairman or secretary. Although the outcome of these special and unusual processes in most cases was quite acceptable, it seems that more hesitation in some of the cases would have been prudent. From my own involvement of some 20 years on the committee as ordinary or adjunct member since 1973, I judge that cases of the kind discussed would be very unlikely to occur today. The interesting question is whether the award to Lederberg in the same year as his first nomination will turn out to be the last of its kind. The only way to learn to know is patience — and longevity — and to follow how, year by year, the secrets of the rich, unique, real-time evaluations of medical sciences provided by the archives of the Nobel Committee of Physiology or Medicine at the Karolinska Institute are unveiled.

Coda — Sometimes it takes time

The theme of this chapter has been various situations when the committee has aimed at a speedy recognition of a discovery. The discovery by James D. Watson and Francis H. C. Crick in 1953 that DNA has a double-helix structure has been highlighted as one of the major discoveries, if not *the* major discovery in biology during the previous century. It is frequently compared to Charles Darwin's publication in 1859 of the *Origin of Species*. Still it took a surprisingly long time before the presentation of the DNA structure was nominated for and recognized by a Nobel Prize. Certain central dogmas are difficult to change and this is well reflected in the relative emphasis put on proteins and nucleic acids in the evaluation of fundamental biological phenomena. The next chapter will discuss how the emphasis has shifted back and forth between these two dominating categories of biomolecules.

Chapter 7

Nobel Prizes and Nucleic Acids:
A Drama in Five Acts

By 1959 there had still been no nominations for a Nobel Prize for the discovery of the double-helix structure of DNA, neither in Physiology or Medicine, nor in Chemistry. One might have expected this revolutionary finding, providing solid support for a role of nucleic acids as carriers of the genetic information in Nature, to have been highlighted by the invited nominees. Jacques Monod, one of the three recipients of the 1965 Nobel Prize in Physiology or Medicine (Chapter 3) lists four major milestones in the advance of biology in his book *Chance and Necessity*[1]. These are Charles R. Darwin's description of the principles of evolution, J. Gregor Mendel's identification of the rules of inheritance, Osvald T. Avery's early identification of DNA as a carrier of genetic information and James D. Watson's and Francis H. C. Crick's presentation of the nature of double-stranded DNA. However, it took time before the central role of DNA was finally accepted by the scientific community. The decisive discoveries leading to this acceptance have been described in an excellent way in a number of books[2-8]. This chapter does not intend to repeat a presentation of the major events. Instead it will highlight the phases of transition in our emphasis on either nucleic acids or protein in our growing understanding of central phenomena of functions of genes. Nobel archive materials were used for the examination of the critical events until 1959. Developments hereafter are presented in a more summary form.

The first prize in Physiology or Medicine to be analyzed is the one given in 1910 to Albrecht Kossel for "… his work on proteins, including the nucleic substances." His contributions can be seen as the climax of some 50 years during which the importance of nucleic acids had a central role in discussions of the nature of hereditary material, at least comparable to that of proteins. Hereafter proteins came to dominate as illustrated in Chapter 3 in the discussion of Stanley's presentation of an infectious plant virus protein recognized by a shared prize in Chemistry in 1946. As developed in the same chapter the progressively growing insight into the nature of viruses was contingent upon new data clarifying the central role of nucleic acids. Critical observations were made both in studies of plant viruses and bacteriophages. It was during the 1950s that the most fundamental shift in our understanding of the chemical nature of the genetic material occurred. These critical events will be evaluated by reviewing the Nobel archives at both the Royal Swedish Academy of Sciences and the Karolinska Institute during that decade. This will be the major emphasis in this chapter.

The missed opportunity to award the prize to Avery will be revisited, although this has already been discussed[9]. Prizes for studies of nucleic acids during the time period to be considered are the one in Chemistry in 1957 to Alexander R. Todd and the two prizes in Physiology or Medicine in 1958 and 1959. The 1958 prize to Beadle, Tatum and Lederberg was briefly referred to in Chapter 3 and in the case of Lederberg discussed more extensively in Chapter 6. The prize in 1959 to Severo Ochoa and Arthur Kornberg finished a decade of fundamental evolution of our understanding of the structure and functions of nucleic acids. This phase of development was sealed by the recognition of the discovery of the structure of DNA by Watson and Crick by the 1962 prize given to them and to Maurice H. F. Wilkins (Figure 1). The prize motivation was "for their discoveries concerning the molecular structure of nucleic acids and its significance for the transfer of information in living material." The circumstances of the latter award remain for future evaluations of archive materials. The discoverers' own descriptions of the events are well known. Watson's controversial book *The Double Helix*[10], originally intended to be called "Honest Jim," became famous overnight. Crick's version of the emergence of the discovery, the book *What Mad Pursuit*[11], is interesting in its own right. It is said that Crick originally intended to use a more Shakespearean title, "The Taming of a Screw." Also Wilkins has given his own version of the discovery in his autobiography *The Third Man of the Double Helix*[12].

Once DNA was back on stage in a leading role it became important to understand the universal language of Nature. It did not take long before the genetic code was cracked and celebrated by a Nobel Prize in Physiology or Medicine in 1968. Interestingly the opportunities to read the books of life shifted the focus back to proteins. Reading the complete nucleotide sequence of

Crick receiving his Nobel Prize from the hands of His Majesty King Gustaf VI Adolf. Watson and Wilkins wait in the background for their prizes. [© Scanpix Sweden AB.]

the hereditary material of a certain species allowed a deduction of the number of proteins that it could specify, and their general characteristics. However, further insights into the significance of nucleotide sequences have made it clear that in the initial enthusiasm for the deduction of protein structures the importance of sequences that code for the second form of nucleic acids, RNA, but do *not* result in formation of proteins, had been overlooked. Important discoveries of new kinds of RNA with unique functions were recognized by Nobel Prizes in 1989 and 2006. These advances opened new fields of research and put nucleic acids, in particular RNA, back on stage. Thus until today one can discern five acts in the drama highlighting the main actors in the play of life, the nucleic acids and the proteins. Nucleic acids have dominated the scene in the first, middle and last act, whereas proteins took center stage in acts two and four.

The chemistry of the nucleus

Friedrich Miescher (1844–1895) was a shy and modest person[13,14]. When he finished his medical studies in Basel in 1868 he took an unexpected course

Friedrich Miescher (1844–1895), the discoverer of DNA. [Courtesy of Friedrich Miescher Institute, Basel, Switzerland.]

in his professional life. Because he had a hearing impairment he decided not to go into clinical medicine but to become a researcher in biochemistry and physiology. He went to study at Felix Hoppe-Seyler's laboratory in Tübingen, Germany. The laboratory was famous for the studies of hemoglobin and its binding of oxygen, but this is not what Miescher wanted to do. He wanted to study the chemistry of the cell *nucleus*. At the time it was appreciated that the remaining part of the cell was composed of proteins, but what were nuclei composed of? This part of the cell had different staining properties than the surrounding part, the *cytoplasm*. He first tried to purify nuclei from white blood cells recovered from pus in bandages from wounds. After certain efforts he managed to collect reasonably pure cells. He then tested different techniques to remove the cytoplasm. Eventually he settled on a method of suspending the cells with warm alcohol to eliminate lipids and then remove proteins by digestion with crude enzyme preparations.

The purified preparation of nuclei was then extracted with an alkaline solution and hereafter acidified. This gave him a precipitate that he called *nuclein*. It was uniquely rich in phosphorus. It took time for Miescher to get his data published, since his mentor, Hoppe-Seyler, first wanted to check that they were correct. When they were published in 1871 he returned, after a few years' break, to his studies of nuclein. He cleverly chose to use salmon sperm, containing cells composed predominantly of nuclei, as a starting material. He was able to show that the main components of these cells were his nuclein and a protein he called protamin. The two components were separated and the nuclein isolated by the procedure contained 9.6% phosphorus, was free of sulphur and contained no detectable protein. The name *nucleic acid* for this kind of preparation was conferred in 1889 by Richard Altmann, a German pathologist and cell biologist, but at that time Miescher had, for unknown reasons, given up his studies of the substance, the discovery of which he pioneered. When Miescher died of tuberculosis in 1895 the interest in his substance had increased. However it was not primarily chemists that were stimulated by his discovery, but also cell biologists.

It was the German biologist and founder of cytogenetics, Walter Flemming, who in 1882 in his book *Cell Substance, Nucleus and Cell Division* stated that Miescher's nuclein was responsible for the unique staining characteristics of the nucleus. A few years later the phenomenon of cell division, *mitosis*, was elucidated. In association with this phenomenon new kinds of structures were identified, the *chromosomes,* from Greek *chromo*, stainable, and *soma*, body. The chromosomes became visible and performed a choreographed dance during mitosis. In 1885 the famous German zoologist Oscar Hertwig wrote "nuclein is the substance responsible not only for fertilization, but also for the transmission of hereditary characteristics." Paradoxically Miescher himself rejected the proposal that his nuclein had a role in genetic phenomena.

Enter Albrecht Kossel

Kossel (1853–1927) had a good portion of the self-confidence that Miescher lacked[14]. He started his medical studies in Strasbourg, where he had the leaders of organic chemistry of the time, Felix Hoppe-Seyler and Emil Fischer, as inspiring teachers. After having finished his medical studies in his home town of Rostock he returned to the laboratory of the former scientist. When Hoppe-Seyler had checked Miescher's findings of nuclein, before its eventual publication in 1871, he went on to look for nuclein in other materials. He found that it was present in yeast. This observation was criticized by some influential chemists of the time, but Kossel in his first studies was able to confirm Hoppe-Seyler's findings. This set him on the track

Albrecht Kossel (1853–1927), recipient of the Nobel Prize in Physiology or Medicine 1910. [From *Les Prix Nobel 1910.*]

of extensive and life-long studies of the chemical characteristics of nuclein, which laid the foundation for our present-day understanding of nucleic acids. It was this comprehensive contribution that was the core motivation for his Nobel Prize in Physiology or Medicine in 1910. His aim was to characterize the chemistry of the cell, especially its nuclear part. In this work he made many fundamental contributions to our understanding of the chemical nature of its nuclein-protein complex.

The essential advances in our growing insight into the nature of nucleic acids can be briefly described as follows. The backbone of their chain-like structure is sugars linked by phosphate groups. In the early 1890s it was demonstrated by Kossel and independently by the Swedish biochemist Olof

Hammarsten, that there were two kinds of nucleic acids, one of the kind found in thymus and another, first demonstrated in yeast. Later on it was demonstrated that all cells contain both kinds of nucleic acids. The difference between the two kinds of nucleic acids resided in the sugar moiety, which in the case of thymic nucleic acid—DNA—was shown much later by one of Kossel's students, Phoebus Levene, to be *deoxyribos* and in the case of yeast nucleic acid—RNA—as shown earlier *ribos*. Four nucleotide bases were found to be connected to the phosphor-(deoxy)ribos backbone. Two of them, *adenine* and *guanine*, were identified not by Kossel but by Fischer. He characterized them as a part of his groundbreaking studies of carbohydrates, for which he received the second Nobel Prize in Chemistry. He referred to them collectively as *purines*. The two purine bases are identical in DNA and RNA, and this is the case also for one of the other two bases, cytosine. But the fourth of the bases shows a difference. Kossel and collaborators first demonstrated that it is *thymine* in DNA, the thymus nucleic acid, and later it was shown, by Levene again, that RNA, yeast nucleic acid, instead contains *uracil*. A common name for the latter three bases is *pyrimidines*. The four letters (with thymine in DNA corresponding to uracil in RNA) used in the universal language of Nature are often abbreviated to A—adenine, G—guanine, C—cytosine and T (U) for thymine (uracil).

Kossel and collaborators also made important studies of the basic proteins, the protamines and histones, associated with the nucleic acids. Overall he had a very successful professional life. After his productive sojourn with Hoppe-Seyler he became a professor in Berlin and after twelve years in this Prussian city, he moved for six years to Marburg, after which he settled in Heidelberg to hold the prestigious chair of physiology at the university. Throughout his life he continued to develop high quality science. In contrast to Miescher, who was never recognized as a pioneer, Kossel was invested with many honors. Three years before his death, the Eleventh Physiological Congress in Edinburgh, the conference where the dramatic discovery of insulin was first presented (Chapter 6), paid tribute to him. On the same occasion Kossel, who was not overtly "patriotic," was very pleased to receive an honorary doctorate in Edinburgh, a city in a former enemy country during World War I. Kossel believed in the universality of science. His most important recognition was the Nobel Prize in Physiology or Medicine in 1910.

Kossel was nominated six times starting in 1902 for a prize in Physiology or Medicine. His work was reviewed in 1903, 1904 and 1910. The reviewer was Karl A. H. Mörner, professor of chemistry and pharmacy at the Karolinska Institute.

Mörner was also Vice-Chancellor of the Institute from 1898 and until his death in 1917. We met him in Chapter 1, as the chairman and secretary of the Nobel Committee for its first 17 years of existence and thus a very central person in matters relating to the early development of the prize. The 1904 nomination by Gustaf Hüfner, a German chemist from Tübingen, highlighted Kossel's studies of nucleic acids and their associated proteins, protamine and histone. He emphasized that it is only by Kossel's fundamental contributions that the significance of Miescher's original identification of nuclein has become fully understood. Mörner made a thorough review and concluded that Kossel was worthy of a prize. In 1910, Mörner made another long and comprehensive review of Kossel's contributions. Most of the review concerned Kossel's work on nucleic acids and to some extent also on the associated basic proteins. In

Karl Mörner (1854–1917), Vice-Chancellor of the Karolinska Institute (1898–1917) and chairman of the Nobel Committee (1901–1917). [Courtesy of the Karolinska Institute.]

the review, the association between the two led Mörner, somewhat confusingly to a modern reader, to refer to them as protein ("egg white") compounds. The identification of the central pyrimidin compounds thymidin and cytosine was highlighted and separately the characterization of the nucleic acid associated proteins was described. It was the terminology of the day that underlies the prize motivation "… his work on proteins including nucleic substances."

In his introduction at the prize ceremony Mörner emphasized why a chemist was given a prize in Physiology or Medicine[15]. The reason was that he was involved in *physiological* chemistry. Kossel's work expanded our understanding of how cells function, which is a prerequisite for gaining insights into how organs and the whole of our body function. Mörner boasted that, in the future, chemistry might allow an understanding of all fundamental mechanisms of life except perhaps our capacity for thinking and associated expressions of life. As we shall see, modern biologists and biochemists also consider such expressions, the human consciousness and memory, as a challenge for future research. In his speech Mörner further stated "… that these nucleic acids by their presence in the cells and their relation to the protein bodies found there, certainly possess *a great biological significance* … ." However, it would take almost 50 years until the central function of DNA as the genetic material since the dawn of evolution became fully accepted by the scientific community.

The eclipse of nucleic acids

The belief that nuclein played a role in fertilization and for transmission of hereditary characteristics was acceptable to the scientific community only for a few decades. This can be followed in the sequential editions of a textbook by a highly respected biologist at the time, Edmund B. Wilson. The title of the book was *The Cell in Development and Inheritance*[16]. In the first edition of the book (1896) it is stated "… the chromatin is to be regarded as the physical basis of inheritance. Now chromatin is known to be closely similar to, if not identical with, a substance known as nuclein … , which analysis shows to be a tolerably definite chemical compound of nucleic acid (a complex organic acid rich in phosphorus) and albumin. And thus we reach the remarkable conclusion that inheritance may, perhaps, be effected by the physical transmission of a particular compound from a parent to offspring." In the third edition of the book which appeared 29 years later the interpretation of the role of nucleic acids has changed dramatically. Wilson then stated: "These facts offer conclusive proof that the individuality and genetic continuity of chromosomes do not depend on the persistence of chromatin in the older sense." The problem was that it was not possible to identify chromosomes by staining in resting, non-dividing cells. This led to the remark "seems to indicate a progressive accumulation of protein components and a giving up, or even complete loss, of nuclein." It took a long time until DNA again appeared in the limelight of the dancing genes of life.

Many developments in the understanding of proteins demonstrate their rich versatility, made possible by an endless number of combinations of the twenty different amino acids that have been identified. Proteins could form large molecules and as discussed in Chapter 3 it was proposed by Stanley that a protein represented the complete entity of a plant virus. By way of contrast nucleic acids were considered as simpler and monotonous molecules. One of Kossel's heirs, Levene, continued to make important contributions to the description of the structure of nucleic acids, but he also came to markedly hold back the development of an understanding of their biological role by a misleading hypothesis. His conclusion that the four bases appeared in equal proportions in biological materials derived from two misconceptions. He did not appreciate that the nucleic acid preparations he studied were mixtures of molecules, which might even out possible differences in the representation of bases in the various categories of these. Furthermore he did not appreciate that nucleic acids, like proteins, could form large molecules. However, in the mid-1930s data started to appear showing that nucleic acids could also be very large.

It took until the end of the 1940s before Levene's canonical principle of equivalent representation of the four bases was challenged. This was done by Erwin Chargraff. In the mid-1930s he had moved from Vienna to become a biochemist at Columbia University in New York. There he developed an impressive career. In his studies of nucleic acids he made two crucial observations. One was that the relative occurrence of the four nucleotides varied depending upon the material studied. Thus Levene was wrong. The other important finding was that there were always equal proportions of guanine and cytosine and of adenine and thymine bases, respectively. Chargraff was in Stockholm in the late 1940s lecturing about his results which impressed many of the younger scientists but they did not convince other more senior scientists of the possible importance of DNA. It was his demonstration that the four bases seemed to be pair-wise complementary that set the stage for Watson's and Crick's demonstration of the structure of DNA in 1953. Almost ten years before this momentous discovery highly controversial results from studies of bacteria causing pneumonia had been published from the Rockefeller Institute.

The milestone discovery by Avery

In 1944 a paper was published in the highly respected *Journal of Experimental Medicine*, by Avery and collaborators[17]. It concerned isolation of a substance that could transfer surface properties between strains of *Pneumococcus* Type III, a bacterium that can cause pneumonia. Different strains of this bacterium were covered by various sugars at their surface which made them appear rough or smooth and gave them different immunological properties. In earlier experiments by others it had been found that the property of one strain could be transferred to another even if the first donor bacterium lacked capacity to replicate. The phenomenon was called *transformation*. The aim of Avery's work was to determine the chemical nature of the transforming substance. The results that he and his co-workers obtained were surprising and controversial. Part of the summary of their publication read "… indicate that, within the limits of the methods, the active fraction contains no reactive polysaccharide and consists principally, if not solely, of a highly polymerized, viscous form of deoxyribonucleic acid." The latter part of the discussion of the article presents three possible qualities of the transforming substance. It is likened to a gene, compared to a virus and finally to a possible "transmissible mutagen."

Oswald Avery (1877–1955), the first scientist to show that DNA represents the genetic material. [Reproduced by permission from the Rockefeller University Press.]

There have been many discussions about the reluctance of the scientific community to accept the message of the article[18–21]. Levene's tetranucleotide hypothesis still prevailed and furthermore it was easy to argue that the preparation might still contain some proteins. Also the preparation of the DNA-degrading enzyme used to control the specificity of the material might have been contaminated by other enzymes. One of the scientists most hostile to Avery's and his collaborators' proposal was Alfred Mirsky, a highly influential scientist, also at the Rockefeller Institute. The change in attitude among leading scientists did not begin to occur until the beginning of the 1950s. There was some confirmation of Avery's work already in 1947, when André Boivin reported that DNA could induce specific mutations in *E. coli*, a favorite in experimental work. In further investigations of pneumococci, Rollin Hotchkiss in 1951 demonstrated transfer of resistance to penicillin by use of DNA and at about the same time Hattie Alexander and Grace Leidy also described transformation by DNA in another bacterium, *Haemophilus influenzae.*

The significance of Avery's transformation experiments was widely discussed in the early 1950s. The considerations of the possible significance of DNA as carrier of genetic information was influenced by the outcome of the Alfred Hershey and Martha Chase experiment with bacteriophages in 1952 (Chapter 3). In 1951 Hershey received a letter from Roger Herriott[5] with a visionary speculation: "I've been thinking — and perhaps you have, too — that the virus may act like a little hypodermic needle full of *transforming principles*; that the virus as such never enters the cell; that only the tail contacts the host and perhaps enzymatically cuts a small hole through the outer membrane and then the nucleic acid of the virus flows into the cell."

In further studies four years later Gierer, Schramm and Fraenkel-Conrat demonstrated that RNA isolated from tobacco mosaic virus was infectious on its own (Chapter 3). How did these developments in the field, including the elucidation of the double-helix structure of DNA in 1953[22], influence the Nobel

Committees at the Royal Swedish Academy of Sciences and the Karolinska Institute? When did they finally accept that nucleic acids are the universal carriers of genetic information?

The growing insight of the Committee for Chemistry

Among the first 50 Nobel Prizes in Chemistry there is a large number identifying advances in the field of organic compounds and biochemistry[23]. Impressive new insights into the structure and synthesis of molecules important in oxygenation and photosynthesis — hemoglobin and chlorophyll — and signaling in the multicellular organism — hormones and vitamins — were recognized by prizes. With the exception of the prize to Kossel very little attention had been given to molecules involved in storing and transmitting genetic information. The prize awarded to Fischer in 1902 included description of the two purines represented in nucleic acids, but the award was primarily for the foundations he had laid for carbohydrate chemistry, as already mentioned. It was not until the end of the 1940s that the committee became seriously involved in the structure of nucleic acid components and their interaction. It may be said that the prize awarded to Stanley in 1946 for his crystallization of an infectious virus protein (Chapter 3) was a recognition of a form of expression of genetic information, but this is not the way it was viewed, even much later[23]. As already mentioned in Chapter 3 (page 92) Westgren, the long-term chairman of the Committee for Chemistry as well as permanent secretary of the Royal Swedish Academy of Sciences for most of this time, as late as 1970 had summarized that viruses were biological entities in the borderland between living and dead matter. However, in all fairness it should be added that, before this statement, he said with reference to the virus particle: "… it has since proved to contain nucleic acid as an important ingredient." During Westgren's reign there was almost no rotation of members of the Nobel Committee. Several members like Svedberg, Tiselius (photos in Chapter 3) and Arne Fredga, a professor of organic chemistry at Uppsala University, remained on the committee essentially throughout this time. Was this a strength or a weakness? How did they come to accept the revolutionary new view of the role of nucleic acids?

In 1949 Todd was nominated for the first time by A. Killen Macbeth, Australia, "for his investigations of biologically important products and

Arne Fredga (1902–1992), member of the Nobel Committee for Chemistry (1944–1976). [Courtesy of the Royal Swedish Academy of Sciences.]

especially the nucleic acids and nucleotides, and the coloring matters derived from aphids and other insects." He was then nominated from 1952 and onwards, often by several proposers, every year until 1957 when he received the prize. Already in the first year of nomination Fredga made a thorough evaluation. He praised Todd's work on nucleotides but noted that much remained to be known. The recommendation to the committee was to continue to pay attention to the findings, but to delay taking a decision on rewarding them with a prize. Fredga followed up his investigations of Todd in 1953, 1956 and 1957. Each time he noted that the work had advanced and become more eligible for a prize. In the 1953 evaluation Fredga cited Todd, stating "... nucleic acids are from a biological standpoint as important as are proteins." From a chemical perspective his impressive achievements were considered appropriate for an award in 1957. Fredga wrote: "A very important stage has now been covered and the principles for the building of nucleic acids now seem to be as well clarified as those of the polysaccharides and proteins. Nucleic acids have probably provided a larger challenge than the other natural products mentioned ... ."

There are only a few references to the possible role of nucleic acids in the transfer of genetic information in the archives regarding the award of the prize to Todd. In a nomination in 1955 it was briefly mentioned: "... can one fully appreciate the brilliant achievement of his school and himself. They have stimulated a great amount of work outside, and such physical pictures of DNA as Watson and Crick have drawn, rest on the structural theories Todd has developed." However, one paragraph in particular in Fredga's final evaluation is highly significant. It reads:

As is already known, different researchers have on the basis of Todd's building principles developed very interesting theories about the conformation of deoxyribonucleic acids. The most spectacular and perspective-rich model emanates from Watson and Crick (reference), which proposes a double helix, thus two chains with a common axis. The model explains among other things that certain nitrogen bases appear to be represented in equimolar concentrations (for example adenine/thymidine and guanine/cytosine). This (proposal) makes it possible to explain how a chain molecule of this kind can reproduce. Since the deoxyribonucleic acids have a role in chromosomes one is dealing in this case with very central questions of natural sciences. Todd said in 1955 that it "has passed beyond the state of conjecture;" the following year he says that there can be little doubt that it gives a true picture in its essentials.

The motivation for the prize to Todd is "for his work on nucleotides and nucleotide co-enzymes." *Nucleotide* is a term used for the complex of a base and the sugar, deoxyribose or ribose, and phosphate. In his Nobel lecture[24] Todd described the many chemical aspects of this work with nucleic acids. As to their importance he said: "Suffice to say that the double helical structure of the DNA molecule adumbrated first by Watson and Crick on these foundations bids fair to open a new era in molecular biology." He emphasized in the lecture that the future lies in developing methods for sequencing and synthesizing nucleic acids.

Lord (Alexander R.) Todd (1907–1997), recipient of the Nobel Prize in Chemistry 1957. [From *Les Prix Nobel 1957*.]

Apparently in 1957 DNA had come of age and its importance was now fully appreciated by the committee. There is in fact additional evidence that this is the case. The first piece of evidence is Tiselius's evaluation of Fraenkel-Conrat and Williams in 1956, already referred to in Chapter 3. The demonstration by these researchers that the RNA extracted from tobacco mosaic virus was infectious was accepted as very strong evidence that this molecule is responsible for carrying the genetic information. The other fact relates to an unexpected nomination for the 1957 prize. In a letter mailed on December 15, 1956 John H. Northrop (Chapter 3) makes a nomination of Avery, Colin M. MacLeod and Maclyn McCarty "in recognition of their discovery of the chemical nature of the pneumococcus transforming principle." The final paragraph reads:

Nucleic acids, then, are the stuff of life, for which so many men have searched for so long. This, I believe, will come to be considered as one of the greatest of all chemical and biological discoveries.

There are two things about this letter that are surprising. The first one is that it comes from Northrop. As described in Chapter 3 he was one of the prime advocates for Stanley's infectious protein and argued that proteins could stimulate the production of more copies of themselves. The second thing to be noticed is that Avery had died on February 20, 1955, which one would assume that Northrop should have known. Interestingly the committee decided to let Tiselius review the three proposed candidates, possibly because MacLeod and McCarty might be candidates on their own. Over two pages he outlined the essence of the experiment by Avery and collaborators and the developments after its publication. Tiselius finished his evaluation in the following way:

> Avery's, MacLeod's and McCarty's discovery thus is of pre-eminent interest for (our) understanding of the chemistry of the genetic elementary processes. Since Avery is no longer alive it seems difficult for me to motivate a prize to the other two. It is indeed difficult for someone not closely acquainted (with the circumstances) to judge the relative contributions by the three scientists in the work that led to the discovery, but Avery appears to have been the leader of the group. The two others have hardly made as large contributions to the field in their later work as some other researchers (R. Hotchkiss, S. Zamenhof). Based on the information I have available I cannot therefore at present support the nomination.

So how did the Nobel Committee for Physiology or Medicine look at Avery's candidature for a prize and when did it realize that DNA was the carrier of genetic information?

Slow appreciation by the Committee for Physiology or Medicine

In contrast to the situation of the committee at the Royal Swedish Academy of Sciences there were members of the Nobel Committee at the Karolinska Institute who had their own engagement in research on nucleic acids. One might think that this would be an advantage, but in practice it worked the other way around. It seems that those who had already had their fingers in the jam-jar containing nucleic acids and not appreciated their sweetness were reluctant to change their opinion as evidence of their tastiness accumulated. Signs of missed opportunities and built-in biases may be two sides of the same coin. There are three actors to be considered in particular.

Einar Hammarsten (1889–1968), professor of chemistry and pharmacy at the Karolinska Institute (1928–1957) and member of the Nobel Committee for Physiology or Medicine almost all that time. [Photo by Ulf Lagerkvist.]

Torbjörn Caspersson (1910–1997), professor of medical cell research at the Karolinska Institute. [Photo by Nils Ringertz.]

They are Einar Hammarsten, professor of chemistry and pharmacy (1928–1957), Torbjörn Caspersson, professor of medical cell research and genetics (1944–1977) and Berndt Malmgren (picture on page 220), professor of bacteriology (1948–1973). Hammarsten was the nephew of the Swedish biochemist Olof Hammarsten, who contributed to the understanding of the existence of two forms of nucleic acids. In his thesis Einar Hammarsten developed techniques for a gentle purification of DNA from thymus. He continued to work with this substance throughout his life and hence knew all the problems in, for example, removing contaminating proteins. In the mid 1930s he performed interesting experiments together with the young Caspersson, who had specialized in the use of the techniques of medical physics, including the ultracentrifuge recently developed by The Svedberg. In these experiments it was found that nucleic acids could have much larger sizes than previously believed. Molecules with a weight approaching one million were found. Like proteins they could form macromolecules.

After this Hammarsten did not make any breakthrough findings, but he was regarded as a very inspiring mentor and teacher to his PhD students. This has been documented by Peter Reichard, who has provided personal details about the failure to award the Nobel Prize to Avery by recourse to the Nobel archives[9], and Ulf Lagerkvist, who dedicated his book *DNA Pioneers and*

Their Legacy[14] to Hammarsten. Both of them were members of his research group before they became professors of biochemistry in (Uppsala) Stockholm and Gothenburg respectively. Reichard had a major influence on the Nobel work at the Karolinska Institute for more than 20 years, starting in the mid 1960s. Decades before that, Hammarsten himself had had a very deep and long-lasting involvement in this work at the Institute. As early as 1929, the year after he was appointed professor at the Institute, he became an ordinary member of the committee. He remained in that position, apart from a few years during the Second World War, until 1947. After a few years as adjunct member (1948–1950) he was back as an ordinary member and crowned his exceptionally long duties as chairman of the committee from 1955 to 1957. He gave the laudation speeches at the prize ceremonies in 1931, 1937, 1953 and 1955, which indicates the importance of his influence.

Caspersson made a rapid career at the Karolinska Institute. In 1944 he became professor of medical cell research and the director of a newly established division of the Nobel Institute for Cell Research at the Institute. He was more of a hands-on researcher than an intellectual philosopher. His particular talent was used to construct optical machinery together with the engineers at his department. The use of the techniques made possible by the machinery had already been outlined in his thesis presented in 1936. Its title, translated from German, is *On the chemical nature of the structures of the nucleus*. It is reminiscent of the goals that Miescher had set for himself some 80 years earlier. Caspersson studied nucleic acids in cells, first by use of a specific technique to stain DNA and then by absorption of monochromatic ultraviolet light. Whereas proteins give a strong absorption at a wavelength of 280 nm, the nucleic acids absorb the light strongly at 260 nm. In the early 1940s he was already proposing, based on his findings, that RNA had a central role in protein synthesis. He was also interested in the structure of chromosomes. By use of staining techniques he demonstrated that the giant chromosomes found in insect larvae contained DNA localized to separate bands. This led to a suggestion that DNA was involved in gene replication. Still, sadly, Caspersson could not take the final decisive step and accept nucleic acids as the genetic material. Long into his career he retained the belief that proteins were the crucial molecules and that his carefully studied nucleic acids only had auxiliary functions.

Caspersson was well aware of all the important developments in the field of genetics and therefore included phages and bacteria in his studies in the late 1940s. He inspired Malmgren to mentor the PhD studies of

Carl-Göran Hedén dealing with the presence of nucleic acids in bacteria infected with phages. No conceptual breakthroughs were made in these studies due to technical limitations and also because of the prevailing biases. Much later, at the end of the 1960s, Caspersson made another important contribution when he and his group, in particular Lore Zech, published a chromosome-banding technique that came to revolutionize medical genetics[25]. He was nominated several times, starting in 1945, for his many contributions, but he never actually received a prize.

The fact that he was a candidate for a prize for obvious reasons restricted his opportunities to participate in the Nobel work during his long tenure at the Institute. This also applies to other strong Swedish candidates for the prize, for example Hugo Theorell, as we shall see. Caspersson's influence on the Nobel work was indirect through colleagues he had worked with or trained. He was in fact an adjunct member of the committee in 1964 and 1965. In spite of his absence from the committee most of the time, he gave the laudatory presentations at the prize ceremonies in 1946 and 1958. In 1959 there is a thorough review of Caspersson's scientific contributions, together with those of Jean Brachet and Alfred Mirsky — the scientist who fired off the major critique of Avery's DNA work — by Holger V. Hydén.

Hydén was a cell biologist who presented his thesis at the Karolinska Institute and then became professor at the School of Medicine in Gothenburg. He was interested in neurobiology and one of his leading hypotheses was that RNA was involved in memory functions. RNA extracted from the brain of rats which had acquired a particular behavior after training was transferred to the brain of untrained rats. It was argued that the latter group of animals assimilated the selected behavior more rapidly than control animals. This finding was later found to be flawed, but it was cited by the prolific and provocative writer Arthur Koestler in his book *The Act of Creation*[26]. This is a rich book on science and science practises. Koestler had his personal view on the culture of science, as is evident from the title of another book of his, *The Call Girls*. He had also a particular interest in parapsychology and when he and his wife had committed simultaneous suicide it was found that he had willed his assets to establish a chair in that subject. There was no university that was willing to accept this donation until eventually an Edinburgh university took it up in order to prove that there was no substance to parapsychology. To go back to Hydén's evaluation of the three candidates one can note that it is very thorough. He gave praise and priority in particular to Caspersson's contributions around 1940. However, the conclusion by the committee following the recommendation by Hydén was that the three candidates should not at the time be considered for a prize.

Caspersson continued to be discussed as a candidate for a prize well into the 1970s. I do not think that I am revealing too much of a secret — considering in particular that the two central actors have been dead for more than ten years — if I mention that there were still discussions at that time as to whether Caspersson's contributions, in particular the revolutionizing banding technique, might be rewarded with a prize. One suggestion was to combine him with another Swede, Albert Levan, a pioneer in plant and cancer cell genetics at Lund University. When I went to high school in the early 1950s, humans still had 48 chromosomes. This information derived from a publication in the 1930s. It remained as an authoritative and uncontested conclusion for a surprisingly long time. Presumably scientists counting chromosomes and ending up with a lower number concluded that they must have lost some of them. The prevailing dogma of the 48 human chromosomes — 24 pairs — was challenged first in 1956 when a visiting scientist in Levan's laboratory, J. Hin Tjio, demonstrated that the true number is only 23 pairs, in total, 46. They were the first cytogeneticists who dared to trust their chromosome count. Interestingly, our closest relatives, the chimpanzees, like other non-human primates, do in fact have 48 chromosomes. It appears that the single human chromosome in pair number 2 is a fusion of two chromosomes in the other primates.

Levan was a very well-qualified scientist and a charming person who also had a talent for playing the cello. I still remember his visit to my laboratory in the early 1960s, where he personally instructed us in how to make preparations of chromosomes by a squashing technique. There was a nice smell of beeswax, used to seal the preparations, in the laboratory afterward. One may ask whether it is an advantage or disadvantage to know candidates for the prize. As a part of the objectivity and impartiality of the selection process it is absolutely essential to keep apart subjective and objective impressions. I trust that in general this is well done. If anything I think the barrier for Swedish Nobel candidates is even higher than for non-Swedish candidates, although I base this on only the one experience when two Swedish candidates, Sune K. Bergström and Bengt I. Samuelsson, made it all the way to a prize in 1982.

Caspersson's co-worker Malmgren became the major evaluator of Avery's candidature for a prize. Malmgren never grew to be one of the more prominent scientists at the Karolinska Institute, but he was an adjunct member of the Nobel Committee in 1952, 1955 and 1958. During these years, in addition to investigating Avery, he reviewed bacteriophage scientists proposed for the prize (Chapter 3). One influential scientist at the Karolinska Institute who could have become involved in the deliberations about Avery's candidature was Theorell. However, since he himself was a strong contender for a prize and was eventually awarded it in Physiology or Medicine in 1955, he could not become involved during the critical years before that. Later on his influence came to be important for the recognition of the rapidly-growing interest in nucleic acids.

Avery was a scientist who was very familiar to the committee. He was nominated for the first time in 1932 and then repeatedly for a total of 46 times. Most of the nominations concerned his contributions to the understanding of the immunological properties of the surface glycoproteins of pneumococcus bacteria. Most often he was nominated together with Michael Heidelberger, also at the Rockefeller Institute. Heidelberger, by the way, is one of the most nominated scientists in the archives and over a remarkably long time, but he never received an award. Heidelberger was still on our list of candidates during the 1970s and I remember noticing that he was nominated for the first time in 1937, the same year in which I was born! He had a very long life and when he turned 100 years old there was a symposium arranged at the Rockefeller University to honor him. On that occasion he apologized, since he was then working only half-time!

Avery's and Heidelberger's important work on the polysaccharide antigens was evaluated a number of times, but it was never considered worthy of a prize. In 1946 the nomination for Avery for the first time cited his work on bacterial transformation. Hammarsten made a three-page review. He was apparently intrigued by the proposed findings and noted that they represented news on a higher level than Avery's earlier findings. He even said "… makes it likely that the activity can be related to a polydeoxynucleotide." Still, he hesitated, with reference to the fact that the proposed activity of DNA was not securely established since the latter molecule has a propensity to associate with other molecules. He recommended the committee to carefully keep up with research in the field but to delay further decisions about a prize. The next time Avery was nominated for his DNA work was in 1952. One nomination, by Herbert S. Gasser, Nobel Laureate in 1944 and director of the Rockefeller Institute, was very thorough. It reviewed Avery's DNA work and put it in a historical perspective. Malmgren made a comprehensive 19-page evaluation. A detailed description of the background of the discovery, including Frank Griffith's original observation of transformation, and the follow-up confirmatory findings were presented. Among other things he discussed Hotchkiss's confirmatory experiments and the demonstration of transformation phenomena with other kinds of bacteria. Towards the end of the review he wrote:

> The proposal presented by Gasser does not concern the original discovery (of the transformation phenomenon?) but the question of the chemical identity of the transforming substance. Gasser nominates Avery, as already presented in the beginning of this review, because of his discovery that nucleic acids of the desoxyribose type carry biological specificity and capacity to orient the biochemical activity of and determine the hereditary properties of microorganisms. If this were correct one would have shown that nucleic acids are endowed with properties which one has not previously documented them as having: the property of biological specificity. Although the reviewer *personally* is prone to believe that the nominator is right, one cannot disregard that there are considerable difficulties in providing secure evidence of this.

The conclusion of the extensive review, in retrospect, is disappointing. It said that it was possible that further research may make Avery's and his collaborators' highly interesting observations worthy of a prize, but that this at the present time was not the case. The spirit of the Caspersson school prevailed and the committee did not take any action.

The nominations for Avery's DNA findings continued from 1953 to 1955. In 1954 one nominator proposed giving half a prize to Beadle and Tatum and the other half to Avery, emphasizing "that Avery's outstanding discovery contributes by focusing attention on the role of nucleic acids in the transference of inheritable characters." It can be noted that the nominations were still registered under the subdiscipline heading of immunology and not genetics. Hammarsten made another brief review. He cited the developments in the field since Malmgren's investigation two years earlier. In particular he discussed Hotchkiss's extended studies of the transformation phenomenon. He mentioned further that the transformation had also now been shown with another bacterium, *N. meningitidis*. He ended his review with the statement: "It can now be concluded that there has been sufficient investigation of the fact that the transformation capacity is carried by DNA and not by a protein." And still he hesitated.

He referred to the fact that the mechanism of the phenomenon was completely unknown and he was also doubtful whether MacLeod and McCarty should be included in a prize. The conclusion by Hammarsten, and the committee he chaired, was that the nominations of Avery did not need to be further discussed that year. And that is the end of the story since Avery died at the beginning of 1955. However, before that event he had received four more nominations for the 1955 prize. One was from Avery's long-term collaborator Heidelberger. Another, particularly interesting, nomination came from Chargraff. This scientist of missed opportunities nominated Hammarsten together with Avery! Hammarsten was nominated for his studies in the 1920s and later leading to the pioneering demonstration that nucleic acids are macro-molecules and Avery for his "truly epochal contribution to our knowledge" and his opening a new field of science — chemical genetics. It is an irony of fate that when this proposal was to be considered, Avery was no longer alive and the other candidate — the main roadblock to the advance of Avery's candidature — was in a situation of being forced eventually to acknowledge that throughout his long professional life he had misjudged the biological significance of his favorite object, the nucleic acids. Of course he could not accept that his own nomination should be discussed in the committee he chaired for another two years until he finally retired.

When Avery published his first article on transformation by DNA he was 66 years old. Had he lived another 15 years this shy bachelor certainly would have received a Nobel Prize. The question is if he would have come personally to collect his prize. He was awarded the prestigious Copley Medal from the Royal Society in the UK in 1946, but he refused to travel to receive it. It was brought to him by the president of the Society, Henry Dale (Chapter 6). A few years later Avery was also recognized in Stockholm by the Swedish Medical Association, which in 1950 selected him for its rare Pasteur medal in gold awarded only once every tenth year. The first medal was awarded to Pasteur himself on his 70th birthday, December 27, 1892. The motivation for the prize specified Avery's engagement in studies of pneumococcus pneumonia and the specific polysaccharide antigens on the surface of the bacteria, but it also included "the discovery that transformation of types of pneumococci is dependent on a specific substance, which is represented by deoxyribonucleic acid." It has not been possible to find out by whom he was nominated for this medal, but the judges who selected Avery for the honor were a group of microbiologists, led by my predecessor Sven Gard, whom we have met many times before in this book (Chapters 3–5). Thus it would seem

Avery's grave. [Photo by Richard Krause.]

that the microbiologists might have been more prone to accept Avery's findings than the biochemists, but their relative influence in the Nobel Committees at that time was apparently too weak to have a decisive impact.

The whole saga of Avery and DNA has been described, not only by Dubos[19] but also by his collaborator McCarty[21,27]. Avery rests in a graveyard in Nashville, Tennessee, and his tombstone is very unpretentious, in accordance with his personality. The picture (on the left) of his tombstone was kindly provided by Richard M. Krause, a former director of the National Institute of Allergy and Infectious Diseases at NIH. He is a specialist in bacterial immunology and appears himself in the photograph above.

Liljestrand included in his posthumous 1970 summary of the Nobel Prizes in Physiology or Medicine[28] the following conclusion about Avery:

> Thus, Avery's discovery in 1944 of DNA as the carrier of heredity represents one of the most important achievements in genetics, and it is to be regretted that he did not receive the Nobel Prize. By the time dissident voices were silenced, he had passed away.

The temperature of engagement increases

Around 1900 Mendel's observations of the elementary laws of heredity were rediscovered. An independent segregation of different inheritable characters was accepted. The need for a term for the elements carrying different traits soon emerged and in 1909 the Danish botanist Wilhelm Johannsen coined the term *gene* derived from a Greek word for birth, descent. There was also a need for a term summarizing the total hereditary information of an organism and a decade later the word *genome* was proposed. This is

Table 1. Discoveries in fundamental genetics and Nobel Prizes in Physiology or Medicine.

Year	Awardees	Prize motivation
1933	Thomas H. Morgan	"for his discoveries concerning the role played by chromosome in heredity"
1946	Hermann J. Muller	"for the discovery of the production of mutations by means of X-ray irradiation"
1958	George W. Beadle Edward L. Tatum Joshua Lederberg	"for their discovery that genes act by regulating definitive chemical events" (½ prize) "for his discoveries concerning genetic recombination and the organization of the genetic material of bacteria" (½ prize)
1983	Barbara McClintock	"for her discovery of mobile genetic elements"

a portmanteau word formed from the two separate terms *gene* and chromo*some*. The major advances in the understanding of the basic genetic mechanisms during the 20th century have been highlighted by Nobel Prizes in Physiology or Medicine (Table 1). It started with a prize to Thomas H. Morgan for his mapping of genes and showing their linkage using the fruit fly, *Drosophila melanogaster*. In his laudatory speech at the prize ceremony Folke Henschen, professor of pathology, emphasized the historical linkage from Darwin via Mendel to the identification in the early 1900s that hereditary properties are carried by the chromosomes. The latter relationship was consolidated by Morgan's findings and is emphasized in the prize motivation "for his discoveries concerning the role played by the chromosome in heredity." Thirteen years after the award to Morgan his collaborator at an earlier stage, Hermann J. Muller, was awarded a prize for the discovery of *mutations* caused by X-irradiation. In his laudatory speech at the prize ceremony on this occasion Caspersson elaborated on these important developments and said that the gene had now been "materialized" and continued "... perhaps a giant molecule of protein character, and, as Muller had first suggested, probably resembling the simpler types of virus which have been dealt with earlier this evening."[29] The latter statement refers to the prize to Stanley for his studies of tobacco mosaic virus (Chapter 3).

The two prizes in Physiology or Medicine in 1958 and 1959 signal the intensified engagement in genetics and the new field *molecular biology* by the now seasoned Nobel Committee at the Karolinska Institute. The term molecular biology had already been introduced in the 1940s by Harry Weaver at the Rockefeller Foundation. Again it was Caspersson who in 1958 addressed the prize recipients[30], George W. Beadle, Edward L. Tatum and Joshua Lederberg.

George Beadle (1903–1989) and Edward Tatum (1909–1975), recipients of one half of the Nobel Prize in Physiology or Medicine 1958. [From *Les Prix Nobel 1958.*]

The half prize to the first time-nominee Lederberg for his revolutionary genetic work with bacteria has already been presented in Chapter 6. Beadle and Tatum received their half of the prize because they were the first to formulate the "one gene one enzyme (protein)" concept.

Beadle was a farmer's son from Wahoo in Nebraska who was the first in his family to continue into higher education[31]. He became involved in studies of the genetics of plants, especially corn, and worked in Rollins A. Emerson's laboratory at Cornell University, Ithaca, New York. In parentheses it can be mentioned that in that laboratory he got to know Barbara McClintock. The recognition of her pioneering contributions took a long time. The committee had to do a lot of homework to understand the significance of her concept of the "jumping genes" and not until 1983 was this tenacious woman awarded a prize. Beadle's and McClintock's ways separated after a few years and Beadle went on from Cornell to California Institute of Technology, Pasadena, California, where he became involved in genetic studies using the fruit fly. However, this system did not allow him to do the studies he intended. He therefore switched to common bread mold, *Neurospora crassa.* Genetic studies with this organism yielded fundamental information about the relationship between the activity of individual genes and different enzyme activities. These findings were developed by Beadle in collaboration with Tatum. At this time they had settled at Stanford University, California. Both these candidates had been nominated for a prize in Physiology or Medicine several times since 1948. They had been subjected to full investigations on four occasions by three reviewers. In the last review, in 1955, Hammarsten reiterated the conclusions of prior investigations that Beadle and Tatum had given a new meaning to the gene concept and were worthy of a prize.

Caspersson, as might be expected, did not mention nucleic acids in his laudatory speech[30] and nor did Beadle in his Nobel lecture[32]. However, it is clear from both Tatum's[33] and Lederberg's lectures (Chapter 6) that nucleic acids have now come of age. Tatum talks about "DNA hereditary changes" and "The relationships between DNA, RNA and enzymes ... ." After 1958 the mechanisms of genetics were ready to be examined in detail by a chemical approach. In this year an important additional piece of evidence for the correctness of the Watson–Crick DNA structure and the consequences it had for conservation of genetic information had been published. In elegant experiments, using heavy isotopes, it was proven that replication of DNA led to that in the product one of the two parental chains was retained and the other one was newly synthesized. It did indeed take time to definitively prove the role of DNA in heredity and it is not correct to state like what Watson has done in one of his books[34] that immediately after the double-helix structure had been published in 1953 "Any doubt as to whether DNA and not protein was the genetic-information-bearing molecule suddenly vanished."

In 1959 the prize was given to Severo Ochoa and Arthur Kornberg for studies of the biological synthesis of ribonucleic acid and deoxyribonucleic acid. Genetics studies then had an emphasis on biochemical events and there was a powerful development of the field of molecular biology. Both these awardees were nominated for prizes in Chemistry as well as in Physiology or Medicine.

Ochoa was nominated for the first time for a prize in Chemistry in 1956. He was evaluated by Karl D. R. Myrbäck, professor of organic chemistry and biochemistry at Stockholm University. Myrbäck had a very positive view of Ochoa's studies of RNA synthesis. Still he recommended that the committee should wait before taking a final position and this was also its decision. The nomination was repeated in the coming years and Myrbäck made another evaluation in 1958 and again in 1959. In the latter year the conclusion was that Ochoa's discoveries now deserved to be recognized by a prize in Chemistry. Kornberg was nominated for this prize for the first time in 1959. The nomination came from a Swedish professor of zoology and cell research at Stockholm University, John Runnström. He concluded his nomination:

Kornberg seems hereby to have brought the problem of the reproduction of the hereditary material within reach for an analysis by chemical methodology and thereby laid the foundation for an understanding of one of the most central problems in biology in exact scientific terms.

Myrbäck also reviewed Kornberg's contributions. He was very impressed by the results presented and already in this first year of nomination stated that they were worthy of a Nobel Prize in Chemistry. However, in 1959 this prize went to Jaroslav Heyrovsky "for his discovery of the polarographic methods of analysis."

Ochoa was first nominated for a prize in Physiology or Medicine in 1954. In the coming years the number of nominations eventually increased to a total of 37. He was evaluated in 1954 and 1956 by Hammarsten, who made complimentary remarks about the work and recommended that the committee should continue to follow the field of RNA synthesis. In 1958 Kornberg was nominated for the first time. The proposal came from Baird Hastings at

The Nobel Committee at the Karolinska Institute, 1959. The persons in the picture are from left Hugo Theorell, recipient of the 1955 Nobel Prize in Physiology or Medicine, Ulf Friberg, Vice-Chancellor of the Institute, Göran Liljestrand, secretary, Ulf von Euler, chairman and forthcoming recipient of the Nobel Prize in Physiology or Medicine 1970, Axel Westman, vice-chairman and professor of obstetrics and gynecology, Carl Gustaf Bernhard, professor of physiology and Berndt Malmgren, professor of bacteriology. [Courtesy of the Karolinska Institute.]

Harvard, Massachusetts, and included Ochoa. At this time Theorell, having received his own prize in 1955, had become heavily involved in the work of the committee. He took control of developments and evaluated the two nominees. The evaluation was very positive, but still he recommended that the committee wait before taking a final decision. He was concerned about the purity of the enzyme used by Kornberg and about the need for so-called primers in Ochoa's system for RNA replication. The following year there was no further nomination of Kornberg. Theorell therefore ensured that Kornberg could be discussed by submitting a nomination himself. This is another case of the kind discussed in Chapter 6 of awardees receiving a prize without any external nomination. In his 1959 evaluation Theorell referred to new data and in the twelve-page report, written at Woods Hole Marine Biology Station, Massachusetts, he concluded that the two candidates were now highly deserving of a prize. Each of them had separately isolated an enzyme involved in replication of nucleic acids from bacteria, in Ochoa's case from *Azotobacter vinelandii* and in Kornberg's case from *Escherichia coli*. Theorell concluded that "The synthesis of nucleic acids is directed by enzymes, which can only act if they have a preformed polynucleotide- or nucleic acid-pattern (primer) to work from."

In 1959 the committee had three ordinary members — von Euler, Westman, Theorell — and three adjunct members — Bernhard, Friberg, Malmgren (see photo on facing page) as listed in *Les Prix Nobel*. However the final protocol was signed by three additional members, who therefore also participated in its work. The additional adjunct members were Sune Bergström, professor of chemistry since 1958 and Nobel Laureate in 1982, Arne Engström, professor of medical physics since 1956 and Klein, professor since 1957. These three scientists illustrate the expansion of the College of Teachers at the Karolinska Institute at the end of the 1950s and the marked addition of qualified evaluators in the Nobel work that this implied. New winds were blowing at the Institute and this was reflected in the increasingly dynamic activities of the committees during the 1960s.

The conclusion of the 1959 committee was that the prize should be awarded to Ochoa and Kornberg. There was only one dissenting voice and that was the vice-chairman Axel E. Westman, professor of obstetrics and gynecology, who instead wanted to see a prize go to Charles Huggins, who eventually received it in 1966 for his work on hormone therapy for certain cancers, and George N. Papanicolaou. The latter scientist was the father of cell diagnostics of smears from the female genital tract for diagnosis of cancer. He was never to receive

Severo Ochoa (1905–1993) and Arthur Kornberg (1918–2007),
recipients of the Nobel Prize in Physiology or Medicine 1959.
[From *Les Prix Nobel 1959.*]

a prize. The College of Teachers agreed with the majority of the committee
and selected Ochoa and Kornberg for the prize. The motivation was "for their
discovery of the mechanism in the biological synthesis of ribonucleic acid and
deoxyribonucleic acid." It is striking that Kornberg received his prize based
only on one external nomination in 1958. Clearly the committee of 1959 was
strongly inclined to award a prize in the field of nucleic acid research.

Theorell gave the laudatory presentation at the prize ceremony[35]. The
earlier prizes awarded to Kossel and to Todd were mentioned in his speech.
He noted that the DNA synthesized by Kornberg was mainly present as the
hereditary substance in chromosomes, but that RNA had other functions,
such as assisting in the synthesis of proteins. In reference to the latter he
mentioned Caspersson's pioneering contributions. He also stated "that the
order of the building blocks in the chains is by no means left to chance, but
on the contrary is planned in detail for each kind of molecule and for each
kind of living organism."

Ochoa was of Spanish origin, but he got his training abroad and also
developed his science outside his home country. He is another example of
the important influence of previous or future Nobel Prize recipients on the
development of a talented young scientist. After receiving a medical degree
at Madrid University he worked from 1929 to 1931 in Meyerhof's laboratory in Berlin (Chapter 6). Following additional scientific training, mostly
in the UK he moved to the US in 1940. There he developed his knowledge
of enzymology at the Washington University Medical School, St. Louis,

Missouri, in the laboratory of Carl and Gerty Cori, the future recipients of the 1947 Nobel Prize in Physiology or Medicine. In 1942 he joined the New York University College of Medicine, where he came to develop the work that led to his receiving the prize.

Kornberg's family originally hailed from Eastern Europe. Little did Ferdinand and Isabella know that by expelling individuals with a non-Christian faith from Spain in 1492 they would thereby plant seeds in that part of Europe from which more than four centuries later many Nobel Laureates would originate. In the post-World War I atmosphere of anti-Semitism, Kornberg's ancestors moved to the US. He grew up in Brooklyn, New York, and received his M.D. from the University of Rochester. His early research experience was gained by working with Ochoa in 1946 and the Coris in 1947. He then developed his own line of research, as chairman of the Department of Microbiology at Washington University School of Medicine in St. Louis, that was to lead to the discovery of DNA polymerase. After the Nobel award he built a world-famous laboratory at Stanford School of Medicine, a Mecca for the training of scientists interested in the chemistry of nucleic acids. He also became a role model as scientist for his own children, as we shall see.

Ochoa's Nobel lecture[36] elucidated the increasing emphasis that is put on RNA, playing a critical intermediate role between DNA and proteins. Kornberg's achievements for which he received the prize turned out to be just the beginning of an impressive development in the understanding of DNA and its replication[37].

From the structure of viruses to the central dogma

During and in particular after 1959 the field of molecular genetics continued to develop with vigor. Following the paradigmatic insight that the genetic information was stored in DNA it became urgent to understand how it was expressed. At this time it became clear that synthesis of DNA was synchronized with cell division. The amount of DNA in a cell doubled prior to the splitting of one cell into two. The only exceptions to this rule were the germline cells, egg cells and sperms. When they were formed they acquired only half the normal amount of DNA. Upon fusion at fertilization the proper amount of DNA was reconstituted. Beyond these basic observations it remained to understand the relative roles of DNA and RNA in relation to protein synthesis and what kind of universal language DNA used. Many Nobel Prizes came to be awarded in this rapidly developing field: first, mostly in Physiology or Medicine (Table 2) but later also

Table 2. Nucleic acids and Nobel Prizes in Physiology or Medicine.

Year	Awardees	Prize motivation
1959	Severo Ochoa Arthur Kornberg	"for their discovery of the mechanisms in the biological synthesis of ribonucleic acid and deoxyribonucleic acid"
1962	Francis H. C. Crick James D. Watson Maurice H. F. Wilkins	"for their discoveries concerning the molecular structure of nucleic acids and its significance for information transfer in living material"
1968	Robert W. Holley H. Gobind Khorana Marshall W. Nirenberg	"for their interpretation of the genetic code and its function in protein synthesis"
1975	David Baltimore Renato Dulbecco Howard Temin	"for their discoveries concerning the interaction between tumour viruses and the genetic material of the cell"
1978	Werner Arber Daniel Nathans Hamilton O. Smith	"for the discovery of restriction enzymes and their application to problems of molecular genetics"
1993	Richard J. Roberts Phillip A. Sharp	"for their discoveries of split genes"
2006	Andrew Z. Fire Craig C. Mello	"for their discovery of RNA interference–gene silencing by double-stranded RNA"

Table 3. Nucleic acids and Nobel Prizes in Chemistry.

Year	Awardees	Prize motivation
1980	Paul Berg	"for his fundamental studies of the biochemistry of nucleic acids, with particular regard to recombinant-DNA" (½ prize)
	Walter Gilbert Frederick Sanger	"for their contributions concerning the determination of base sequences in nucleic acids" (½ prize)
1982	Aaron Klug	"for his development of crystallographic electron microscopy and his structural elucidation of biologically important nucleic acid-protein complexes"
1989	Sidney Altman Thomas R. Cech	"for their discovery of catalytic properties of RNA"
1993	Kary B. Mullis Michael Smith	"for contributions to the developments of methods within DNA-based chemistry" (two ½ prizes)
2006	Roger D. Kornberg	"for his studies of the molecular basis of eukaryotic transcription"
2009	Venkatraman Ramakrishnan Thomas A. Steitz Ada E. Yonath	"for studies of the structure and function of the ribosome"

in Chemistry (Table 3). Since the archives of these prizes are not as yet accessible and furthermore as there are many excellent books dealing with the history of the development of molecular genetics[3,8] I will only give some summarizing remarks in the remaining part of this chapter. Some exciting fields for the future forays of curiosity-driven young scientists will be highlighted.

After the presentation of the structure of DNA in 1953 Watson left Cambridge to work at the California Institute of Technology but he returned to the Cavendish

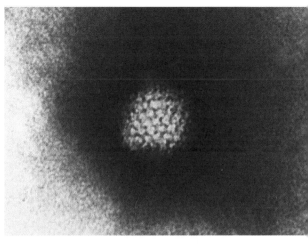

Electron micrograph of an adenovirus particle. The "antennas" projecting from the vertices of the regular icosahedron are used to anchor the virus at the surface of a cell, hereby initiating infection. [Private photo.]

laboratory in Cambridge for the period 1955–1956. At this time he and Crick concerned themselves with the principles of construction of virus particles. They reasoned that the protein shell protecting the nucleic acid of the particles — the transport form of these "infectious genomes" — must be built up of repetitive units forming some kind of symmetrical structure. It was proposed that the protein building blocks had a capacity to self-aggregate into such a regular structure. The visionary postulations of their "Gedanke"— experiments turned out later to be correct.

The full appreciation of the symmetry of the particles was unravelled when a new technique for examining them by electron microscope was introduced in 1959 by Sydney Brenner, who we will soon meet again, and Robert W. Horne. Instead of labeling the proteins with an electron dense stain to make them visible, particles were suspended in a stain of this kind that did not attach to proteins. This "negative" staining method allowed identification of much finer details than previously had been seen. The beautiful symmetry of virus particles or components forming them could be visualized for the first time. The figure above shows an adenovirus particle with its regular icosahedral shape and projections extending from its twelve corners. These projections are used to anchor the particle at the surface of cells, starting infections of them that may lead to respiratory, enteric and other diseases in us humans.

Clarification of the diversity of symmetry of different kinds of virus particles allowed the introduction in 1962 of a practical system for classification

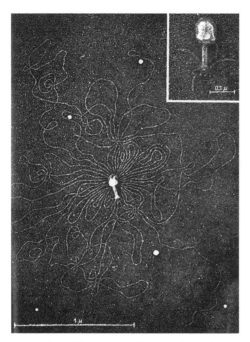

DNA released from T-even bacteriophage. [From Kleinsmidt, A. K., *Biochimica et Biophysica Acta*, 61, 861 (1962).]

of viruses. Three fundamental characteristics of the particles were considered; their content of nucleic acid — either RNA or DNA; symmetry and complexity of the protein cover; and absence or presence of additional membrane-like structure(s). Many different families of viruses that can cause diseases in man and other species could be distinguished. However, they were not families in a true Linnaean sense since the evolutionary origin of viruses still remains to be explained today. It is in fact likely that viruses may have different kinds of origin.

In 1962 further developments in electron microscopic techniques also made it possible for the first time to see the elusive DNA. The figure above shows a bacteriophage that has been subjected to an osmotic shock leading to a release of the DNA contained in its head. In order to visualize the DNA it has been shadowed, at an angle, with a spray of metal. Only two free ends of the coiled molecule can be seen and hence it represents a single long structure. When seeing the picture it is hard to believe that there is space enough for the long DNA in the head of the virus particle. However, the molecule appears much thicker than it is because of the metal shadowing. One obvious question to answer is how the DNA is packed into and released from the head of the particle. In later studies an impressive, miniature molecular motor has been identified in the tail of the bacteriophage. This motor effectively transports the long string-like molecule, piece by piece, into the preformed empty head. When the phage has anchored at the surface of a bacterium to infect it, the reverse process takes place. The DNA is transported through the tail and injected into the cytoplasm, explaining the early surprising findings in the Hershey–Chase experiment (Chapter 3).

Overall the transport and packing of the huge DNA chains into cellular structures too represent a major challenge. Each one of the ten trillion cells in our body contains about two meters of DNA. The total length of our entire DNA would take us many times back and forth from Earth to the moon. Or to take another analogy, if the nuclei of the galaxy of cells in our body had the size of a tennis ball, the total length of the 46 pieces of chromosomal DNA would measure 40 kilometers (24 miles).

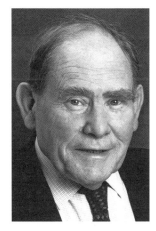

Sydney Brenner, who shared a Nobel Prize in Physiology or Medicine with Robert Horvitz and John Sulston 2002. [From *Les Prix Nobel 2002.*]

After their excursion into the world of virus structures Crick and Watson concentrated on the central questions in information flow between nucleic acids and proteins. Crick was the dominating theoretical molecular biologist in the early 1960s and his activities in the field have been described in detail[8,38]. He contributed many seminal proposals regarding fundamental mechanisms, including intense speculations about the genetic code throughout the later 1950s and he selected this theme for his Nobel lecture[39]. It is in fact striking that neither Crick nor Watson discussed the developments that led to the recognition of the DNA double helix in their Nobel lectures. Only the invisible Wilkins discussed crystallography and DNA. It can be mentioned that Wilkins had declined to be a third author on the famous 1953 *Nature* publication[22], something he was later said to have regretted[34]. Watson in his lecture discussed the developments in the understanding of the functions of RNA[40].

In 1962 a new understanding of the different functions of RNA had emerged and Crick had formulated his famous *central dogma*. This dogma says that the information flow is from DNA to RNA to protein. The two fundamental steps are referred to as *transcription* and *translation*. A special form of RNA, *messenger RNA*[41] carries the linear information of DNA to the protein-synthesizing machinery in the cytoplasm. The latter concept, derived from work by many scientists, was developed by Sydney Brenner, one of the most important scientists among Crick's intellectual sparring partners. Brenner grew up in South Africa but he is another example of a successful scientist and Nobel Laureate, who has a Jewish ethnic background with roots in Eastern Europe. Brenner is a man who enjoys puns and colorful expressions in his lively conversation. Paraphrasing E. M. Forster he has sometimes referred to his lodging at a hotel as "A room with a Jew."

(*Reprinted from Nature*, Vol. 192, No. 4809, pp. 1227–1232, December 30, 1961)

GENERAL NATURE OF THE GENETIC CODE FOR PROTEINS

By Dr. F. H. C. CRICK, F.R.S., LESLIE BARNETT, Dr. S. BRENNER and Dr. R. J. WATTS-TOBIN

Medical Research Council Unit for Molecular Biology, Cavendish Laboratory, Cambridge

THERE is now a mass of indirect evidence which suggests that the amino-acid sequence along the polypeptide chain of a protein is determined by the sequence of the bases along some particular part of the nucleic acid of the genetic material. Since there are twenty common amino-acids found throughout Nature, but only four common bases, it has often been surmised that the sequence of the four bases is in some way a code for the sequence of the amino-acids. In this article we report genetic experiments which, together with the work of others, suggest that the genetic code is of the following general type:

(*a*) A group of three bases (or, less likely, a multiple of three bases) codes one amino-acid.

(*b*) The code is not of the overlapping type (see Fig. 1).

(*c*) The sequence of the bases is read from a fixed starting point. This determines how the long sequences of bases are to be correctly read off as triplets. There are no special 'commas' to show how to select the right triplets. If the starting point is displaced by one base, then the reading into triplets is displaced, and thus becomes incorrect.

(*d*) The code is probably 'degenerate'; that is, in general, one particular amino-acid can be coded by one of several triplets of bases.

The Reading of the Code

The evidence that the genetic code is not overlapping (see Fig. 1) does not come from our work, but from that of Wittmann[1] and of Tsugita and Fraenkel-Conrat[2] on the mutants of tobacco mosaic virus produced by nitrous acid. In an overlapping triplet code, an alteration to one base will in general change three adjacent amino-acids in the polypeptide chain. Their work on the alterations produced in the protein of the virus show that usually only one amino-acid at a time is changed as a result of treating the ribonucleic acid (RNA) of the virus with nitrous acid. In the rarer cases where two amino-acids are

Reprint with Crick's and Brenner's signatures. [Private photo.]

Although Brenner did not receive a prize until 2002 he made many prize-worthy contributions as early as the 1950s and 1960s. He was brought to Cambridge by Crick and they shared an office for some 20 years until Crick moved to the Salk Institute in La Jolla at the end of the 1970s. Their combined minds fostered many new developments. Not only has the 1953 article in *Nature* by Watson and Crick of the structure of DNA become a collector's item, but so have follow-up publications. The picture on page 228 shows a reprint of an article discussing the genetic code published in 1961 with Crick's and Brenner's signatures, which was kindly given to me by Richard A. Lerner, president of the Scripps Research Institute in La Jolla. The two visionary scientists were again reunited for a few years starting in 2000. For a while Crick, Brenner, Leslie Orgel, another very influential molecular biologist during the 1950s and 1960s, and Renato Dulbecco had offices in the same corridor at the Salk Institute. One can assume that the conversations at coffee breaks may not have been only about the results of San Diego Padres' last game. Dulbecco was recognized by a prize in Physiology or Medicine in 1975. This prize recognized the discovery that there are enzymes that transcribe RNA into DNA, *reverse transcriptase,* and their role in RNA viruses causing tumors. This prize was shared with David Baltimore and Howard Temin. Their observation served as a supplement to the central dogma recognizing that information can also flow "backward" from RNA to DNA. However the dogma excludes the flow of genetic information from proteins to nucleic acids.

Crick was always arguing that studies of fundamental phenomena should be pursued in the simplest possible system. Thus if one could study human cancer genes in yeast cells this is what one should do. In that spirit Brenner introduced the experimental use of a relatively simple multicellular system. This was the soil nematode *Caenorhabditis elegans.* It has offered great opportunities to study the development of, for example, the central nervous system. This animal was used in the work rewarded by the 2002 Nobel Prize in Physiology or Medicine, which included Brenner together with H. Robert Horvitz and John E. Sulston "for their discoveries concerning genetic regulation of organ development and programmed cell death." The emergence of the nervous system during embryonic development could be mapped down to the individual cell. The total network of this system including about one thousand cells formed by controled division and programmed cell death — *apoptosis,* a poetic Greek word meaning falling of autumn leaves — could be mapped.

The recipients of Nobel Prizes in Chemistry, Physiology or Medicine and Literature in 1961. From left: Maurice Wilkins (1916–2004), Max Perutz (1914–2002), Francis Crick (1916–2004), John Steinbeck (1902–1968), James Watson and John Kendrew (1917–1997). [© Scanpix Sweden AB.]

Organic crystals unraveled and the code of life deciphered

X-ray crystallography was discovered to be an indispensable means of characterizing the three-dimensional structure of proteins as well as nucleic acids. According to Westgren, the chairman of the Nobel Committee for Chemistry, there were interesting developments in 1962[23]. He wrote in 1970:

> In the early summer of 1962 the medical and chemical Nobel Committees held a joint meeting to discuss possible awards to common prize candidates. The result was that, at the end of that year, five scientists who had applied X-ray diffraction methods to study the vital substances of exceedingly complicated structures were rewarded. Crick, Watson and Wilkins together received the prize in Physiology or Medicine for their research into nucleic acids, while the prize in Chemistry was divided equally between the Englishman John C. Kendrew and the Austrian Max F. Perutz for their research into the structure of globular proteins.

(See above photo.)

Klug refined the technique of diffraction further and could describe in detail the relationship between the RNA and the protein of the rod-shaped particles of tobacco mosaic virus for which he was awarded the 1982 prize in Chemistry. Remarkable further developments of the technique have allowed amazingly detailed descriptions of very complex interactions involving a large number of protein as well as RNA molecules. In 2006, Arthur Kornberg's son Roger became the single recipient of the Nobel Prize in Chemistry for his elucidation of the molecular basis of the transcription apparatus in nucleated cells. Three years later the corresponding findings concerning the second fundamental machinery, the one for translation—the *ribosome*—was rewarded with a prize (Table 3). The prize, again in Chemistry, was given to Venkatraman Ramakrishnan, Thomas A. Steitz and Ada E. Yonath "for their studies of the structure and function of the ribosome."

It is of course a particular event when, as in the case of the Kornbergs, a father and son receive Nobel Prizes, but it is not at all unprecedented. There are in fact six families — Thomson, Curie, Bragg, Bohr, von Euler and Siegbahn —in which parent(s) and offspring have received prizes. In one case the Braggs, father and son, were recognized simultaneously by a prize in Physics "for their services in the analysis of crystal structures by means of X-rays." At the time he was rewarded with his prize the son Lawrence was only 25 years old. He was later to become the head of the Cavendish laboratory in Cambridge where Crick received his training. Another case worth mentioning is that of Iréne Joliot-Curie, who received a Nobel Prize in Chemistry in 1935. Both her father and her mother were Nobel Laureates and the latter was recognized twice, in Physics in 1903 and in Chemistry in 1911.

As early as his 1959 review of virologists studying nucleic acids, Georg Klein (Chapter 3) noted: "Heredity has now become a code construction problem." The question of the nature of the code engaged many of the foremost scientists in the field, including Crick, as mentioned. The definitive identification of the universal genetic code was recognized by a prize in Physiology or Medicine in 1968. It was awarded to Robert W. Holley, H. Gobind Khorana and Marshall W. Nirenberg. Nirenberg had developed a system for protein synthesis outside of cells. His initial unexpected finding was that when he, as a control, fed this system an RNA chain containing only uracil he found a formation of a strange inert and insoluble substance which eventually turned out to be a long protein chain composed only of the amino acid phenylalanine. This allowed the identification of the first code word, UUU gives phenylalanine.

From left: Robert Holley (1922–1993), Gobind Khorana, Marshall Nirenberg (1927–2010), recipients of the Nobel Prize in Physiology or Medicine 1968. [From *Les Prix Nobel 1968.*]

Khorana developed methods for synthesis of RNA molecules containing different predetermined sequences of bases. These could then be used in the protein synthesis system. Holley's contribution was different. He identified and characterized a certain kind of adaptor molecules, *transfer-RNA*, which carried a specific amino acid to the specific site for inclusion in the growing polypeptide, the nascent protein, in the ribosome. The existence of this kind of molecules was first proposed by Crick. The identification of this kind of RNA is fundamental also from a general principle point of view when it comes to formation of transcripts. Its existence implied that that there must be stretches of DNA that code for RNAs, which *do not* in their turn specify a protein. Such non-protein coding RNA has been recognized later to have a much larger role than originally appreciated, as we shall see.

Since the nucleic acids have only four different bases, it was recognized at an early stage that three of them are required to specify each one of the 20 amino acids in the protein chain[42]. A *triplet code* has 64 code words ($4 \times 4 \times 4$) and hence the code is redundant. One amino acid can be specified by more than one triplet, also called a *codon*. The number of code words specifying one amino acid varies and is roughly proportional to the frequency with which it is used in building proteins. The code is comma-less and the reading of a sequence of DNA can in principle start at any point. However there are certain signals defining the beginning and the end of the message for a specific protein. There are special triplets used only as stop codons and a new protein chain starts at AUG, which also codes for the amino acid methionine.

Let me finish this section with a story about the powerful scientific institutions in La Jolla, California, and about Crick. I had the privilege to meet Crick a number of times in that city. He had moved there to work at the Salk Institute in 1977 and completely shifted his field of interest. For many years through the 1980s and most of the 1990s I was a visiting scholar at the Scripps Research Institute. In collaboration with Lerner synthetic peptides were used to map the detailed immunological properties of viral antigens. Lerner has a particular interest in his own studies in catalytic antibodies that have the binding specificity of antibodies and at the same time carry enzymatic activity. They have been found to have considerable applications in the field of medicine.

Becoming Permanent Secretary of the Royal Swedish Academy of Sciences in 1997 forced me to relinquish my connection with the Institute, since the Academy carries responsibility for the prizes in Physics and Chemistry and candidates in the latter field might be active at it. Instead I moved across North Torrey Pines Road to a corresponding position at the Neurosciences Institute headed by Gerald M. Edelman. He received the 1972 prize in Physiology or Medicine together with Rodney R. Porter "for their discoveries concerning the chemical structure of antibodies." Interestingly some time after he had received his prize he completely changed field and went into neurobiology. He founded the Neurosciences Institute in association with the Rockefeller University in 1981 and moved it to La Jolla in 1993. The institute is involved in many different studies, ranging from what behavior Darwin robots can learn, to dreams of the fruit fly. It seems that the remarkably versatile intellect of some laureates in Physiology or Medicine causes them to become attracted to the neurosciences. Their aim becomes to challenge problems which Mörner in 1910 viewed as possibly not approachable by human science[15]. This applies not only to Edelman, but also to another immunologist, Susumi Tonegawa, who received the prize in Physiology or Medicine in 1987 "for his discovery of the genetic principle for the generation of antibody diversity." He has also switched into neurosciences and for a number of years has been the leader of a Center for Neural Circuit Genetics at the Massachusetts Institute of Technology and more recently, in parallel, the RIKEN Brain Science Institute outside Tokyo. In the same spirit Crick spent the last decades of his life researching human consciousness. One can imagine the discussions between the two giants Crick and Edelman at the time when they were both active in this field in La Jolla. They had decidedly different opinions about what the foundation of the human consciousness might be.

Each month there is a meeting in La Jolla of a group of high-profile scientists at the Thursday Club, a gathering somewhat akin to the Moonlight Society in Edinburgh in the late 18th century (Chapter 8). I have been invited to participate in their meetings on a number of occasions as a guest of my good friend Gustaf Arrhenius, a reputable scientist at the Scripps Oceanographic Institute in the field of the origin of life. He is, by the way, a grandson of Svante A. Arrhenius, who had a big influence on the awarding of Nobel Prizes in natural sciences during the first 20 years of their existence (Chapter 6), and who was himself a recipient of the third Nobel Prize in Chemistry. In 2003 I participated in a meeting with the Thursday Club. I had a charming discussion with Crick and when he drove away in his white Mercedes there were two things that came to my mind. One was that I would probably never see him again, since I knew that he was fighting a serious cancer, and the other was that the license plate of his car was highly appropriate. It said ATGC, the four letters of life, and if anyone should have that on his car it was Crick.

Reading the books of life

Frederick Sanger, the two-time Nobel Laureate in Chemistry, 1958 and 1980 (a shared prize). [Courtesy of MRC Laboratory of Molecular Biology.]

Once the universal language of life was deciphered the next step would be to determine how one could read the sequence of bases in a long strain of nucleic acid. This was solved by a remarkable character in the annals of Nobel Prizes, Frederick Sanger. He is the only person who has received two prizes in Chemistry. First he pioneered the development of techniques to determine the sequence of amino acids in a protein. He selected insulin (Chapter 6) and found that it was in fact composed of two chains, one 21 and one 30 amino acids long. The exact position of these 51 different subunits was defined. The two chains were demonstrated to be linked by two disulphide bonds connecting pairs of cystein amino acids. Because of this pioneering work he was awarded the 1958 prize with the motivation "for his work on the structure of protein, especially that of insulin." He continued to work in the fertile Cambridge environment, but moved in 1962 to the new Laboratory of

Molecular Biology established by the Medical Research Council, which came to house such scientific giants as Crick, Brenner, Kendrew, Klug, Perutz and others.

The goal of Sanger's science at this time completely shifted focus. Now he wanted to develop a technique to determine the sequence of bases in a nucleic acid. He selected the DNA from a bacteriophage, φX174. There were many years of painstaking work before the complete sequence had been determined. For a number of years no publications emerged from the work. This is rarely accepted in normally very competitive scientific environments, but in the Cambridge laboratory Sanger — possibly also because of his earlier achievements — had this freedom of action. Eventually he had determined the exact sequence position of the 5,386 bases in the bacteriophage DNA and could deduce the presence of ten genes directing the synthesis of different proteins. This achievement led to his second award, half a prize shared with Walter Gilbert in 1980. The motivation was "for their contributions concerning the determination of base sequences in nucleic acids." A few years later Sanger closed his laboratory and withdrew to tend his garden and build a boat.

The access to a sequencing technique opened the possibility of reading the books of life of ever more complex organisms. A number of supplementary techniques for cutting, cloning and multiplying DNA (Tables 2 and 3) were developed and in the end even very complex genomes could be approached. It became possible to advance from studies of the relatively simple virus genomes, to genomes of bacteria — *prokaryotes*, lacking nuclei — and finally to genomes of nucleated cells, *eukaryotes*, both in single and multicellular forms. Important steps along the way were firstly the determination of the whole genome of a bacterium, *Haemophilus influenzae*[43], and hereafter many similar organisms. A few years later the complete sequence of bases in the DNA from the fruit fly was characterized[44] and eventually the human genome was deciphered in a rough form in 2001 by use of two different approaches. Refined versions of this genome have later been published[45,46].

In order to understand our origin we have been particularly eager to characterize the genomes of our closest relatives. Archeological studies using molecular genetic techniques have recently given some remarkable results. It has become possible to selectively identify small fragments of DNA coming from Neanderthal individuals in more than 30,000-year-old bone fragments heavily contaminated by microbial DNA[47]. The impressive reconstitution of a draft sequence of the genome from our closest evolutionary relative — now extinct since about 30,000 years — allows some amazing comparisons. It can be documented that since the separation of our two lineages of primates some 400,000 years ago 88 major protein-coding changes have occurred. Further examination of these changes may give indications of what makes us human, including our capacity for language. A very tantalizing additional finding is that we humans, who moved out of Africa some 60,000 years ago have 1–4% of Neanderthal genes in our genome, whereas our ancestors who have stayed in Africa have no such DNA[47]!

Each year a rapidly increasing number of genomes representing different species are being characterized. Techniques for sequencing DNA are continually being improved and the consequences of the wealth of ever more rapidly accumulating information are enormous both for the understanding of general biology and for human health. The amount of information that can now be retrieved is essentially endless and as expressed by Brenner in his Nobel lecture[48] we are "… drowning in a sea of data and starving for knowledge."

A sample of sea water contains one million microorganisms and ten million viruses in each milliliter. All the genetic material present in such a sample can be characterized by fragmenting the DNA into suitable pieces and then sequencing all these millions and millions of fragments. Hereafter the cornucopia of fragments, produced by this "shotgun" technique can be combined by use of their overlapping ends into different full genomes or parts thereof. This approach is referred to as *metagenomics* (Chapter 3). It is highly complex and requires qualified competence in the new multidisciplinary field of science called bioinformatics. By this approach it has been possible in two consecutive studies[49,50] to quadruple the number of protein-coding genes identified by man to a total of over twenty million. We are still only in the early stages of exploring the comprehensiveness of the protein universe, in particular as concerns the myriad of invisible forms of life. The function of most of the genes identified has not been defined. They remain a treasure for future insights into the diversity of all the molecular mechanisms exploited by Nature. The new technologies also allow us to study organisms that have never before been identified by man because they cannot be made to replicate in the laboratory. For the first time it has now become possible to make a full inventory of all forms of life. We can finally examine the part of the whole world of life that Darwin could not see.

As mentioned in Chapter 3 we have so far identified only some 6000 "species" of microorganisms, but the true number might be one million times larger. The term species was placed within quotation marks because it has become apparent that in contrast to eukaryotic forms of life with sexual reproduction, individual species of bacteria can show an enormous variation both in composition and size of genomes. Their size can vary markedly, with only a fraction of the genome carrying the fundamental information for the replication of the organism. The rest of the DNA represents genome parts providing supplementary information. This "excess" DNA can be effectively transferred between members of the bacterium species, either on its own or with the help of viruses (transduction, Chapter 6). This wide-ranging *horizontal gene transfer*, with emphasis on survival of the group rather than on the individual, gives particular conditions for the success of a species in the evolutionary competition.

The species concept is also very difficult to apply to viruses. As mentioned, viruses are classified into families according to the structural properties of their transport form, the virus particle. In many cases comparisons of genomes of representatives of different families reveal very little or no evolutionary relationships. This indicates that different kinds of viruses can have separate origins as mentioned. Since viruses are cellular parasites it is tempting to assume that viruses arose after cells. However this is probably not correct. Most likely many of them evolved in the "pre-cellular" phase of evolution. In particular it is likely that viruses using RNA as their genetic material emerged at the time of the RNA world, probably contributing to the early combinatorial evolutionary steps that led to the emergence of progressively more complex forms of primitive life. This pre-cellular world will be discussed further below.

A final word on RNA viruses concerns the inaccuracy of their nucleic acid replication. In comparison with the replication of DNA, which has an impressively high precision, replication of RNA results in about one thousand times more mistakes. Thus no nucleic acid in, for example, one influenza or polio virus particle is identical to that of another. Replication of an RNA virus leads to the appearance of a cloud of viruses with slightly different genomes, referred to as *quasi* — similar but not identical — species. This phenomenon also implies that RNA viruses have a very high mutation rate, something causing major problems in the development and use of drugs to control RNA virus infections. Resistant virus strains easily emerge. The only way of controling this problem is to use three or more kinds of drugs with different points of attack.

Writing the books of life

Ochoa finished his Nobel lecture[36] with the following visionary statement: "Since RNA is the genetic material of some viruses, the work reviewed in this lecture may help to pave the way for the artificial synthesis of biologically active viral RNA and the synthesis of viruses. These particles are at the threshold of life and appear to hold the clue to a better understanding of some of its fundamental principles." This has been successfully done although it took into this century before the required technology was available. In 2002 the whole RNA genome of poliovirus was synthesized and shown to be functional[51] and a few years later the RNA of influenza virus that caused the dramatic 1917–1918 pandemic was synthesized and shown to be pathogenic in animals[52]. The re-creation of the virus that in 1918–1919 killed more than 50 million people raised a discussion of potential risks of publishing such data. Could malicious individuals use this information for the purpose of bioterrorism? Wisely, it was decided that the data should be published. The discussion of the potential dual use of scientific data is as old as science itself. Ever since humans invented, for example, the knife, we have been aware of its dual uses. Hence there is no reason to forbid any form of science, but it is important to ensure that there are ethical rules which guarantee that new information is used and not misused.

The infectivity of isolated viral RNA was demonstrated in 1956 (Chapter 3) but it took five more years before DNA isolated from a virus was also shown to be infectious. This was done with nucleic acid prepared from the same bacteriophage, φX174, that Sanger later sequenced. In 1967 it was demonstrated that new phage DNA could be synthesized by DNA polymerase using an intact genome as a template. The achievement was hailed as "life in the test tube." In this experiment the researchers used Kornberg's polymerase preparations, but there was a need for one more enzyme that could make a circle out of the newly-formed linear double helix molecules, a prerequisite for them to become infectious. In fact there were also some problems with Kornberg's original enzyme. A British geneticist John Cairns showed in 1969 that *E. coli* that lacked that enzyme could survive, but had some problems in repairing its DNA after irradiation with ultraviolet light. If the enzyme Kornberg had discovered had had the central role in DNA replication that it was assumed to have, the bacterium with the mutant would not have survived. It took many more discoveries before eventually the complexity of DNA replication was understood[37]. Another of Kornberg's gifted sons, Tom, managed to demonstrate two more DNA polymerases from *E. coli*. They were referred to as DNA polymerase II and III, since the original enzyme, the discovery of which was rewarded with the Nobel Prize, was labeled I. It then became clear that it was enzyme III that is the true DNA replicating enzyme, whereas the other two enzymes had auxiliary functions. The fact that the discovery of DNA replication was not complete in 1959 does not detract from the achievement leading to the prize. Most of the further developments in the field did in fact occur in Kornberg's fertile laboratory at Stanford.

As sequences of genomes started to be identified it became possible to take a new approach to synthesis of nucleic acids. By linking up, in a predetermined order, bases coupled with the proper background structures into nucleotides, synthetic molecules carrying specific genetic information could be produced. In 2003 it was possible for Hamilton O. Smith in collaboration with Clyde A. Hutchinson III and J. Craig Venter to reconstitute the whole genome of φX174 from synthetic polynucleotides and show that the DNA produced had the capacity to infect bacteria and produce new phages[53]. The strain of virus used in the study had come from Sanger's laboratory, where Hutchinson had worked previously. The ambitions to produce increasingly larger genomes with high fidelity have grown with time. The goal of Venter's research group has for a long time been to produce the whole genome of a small mycoplasma. On May 21, 2010 the team could publish the successful outcome of their efforts[54]. Starting from individual nucleotides it was possible to assemble about 1.08 million base pair long DNA molecule representing the complete genome of *Mycoplasma mycoides*. This synthetic genome was successfully transplanted into cells of a related bacterium, *Mycoplasma capricolum*, whose own genome had been incapacitated. The introduced genome took complete control of the derivative cell. This achievement was not only a major technical feat but it was also a discovery. The result obtained showed that in prokaryote cells the whole metabolic machinery is controlled by the DNA genome. It remains to see the potential need for supplementary information molecules in the much more complex nucleated cell. At least the result of the first two out of the totally 3.8 billion years of evolution of life on Earth has been recapitulated. Still it is not a question of creating new life. The achievement derived from first reading a book of life provided by Nature and then rewrite it in the laboratory. Since it is now possible to recreate the whole genome of a bacterium opportunities for genetic engineering potentially allowing the production under controled conditions of many interesting products have become accessible. It is striking that DNA serving as software has the capacity under certain conditions to build its own hardware.

The name Smith appears repeatedly in the evolution of techniques to characterize and synthesize DNA. It was he who worked in tandem with Venter in the pioneering sequencing of bacterial, insect and human DNA and he is also a lead figure in the synthesis of ever larger DNA molecules. When

Hamilton Smith, recipient of a shared Nobel Prize in Physiology or Medicine 1978. [Private photo.]

Smith was awarded his Nobel Prize in Physiology or Medicine in 1978 he got a telephone call from his mother. She said, "Ham, I just heard over the radio that someone with your name also working at Johns Hopkins Medical School has received the Nobel Prize. Are there two of you with the same name?" The answer was "No, mother, it's me." Not only was Smith's capacity as a scientist unknown to his mother but also largely to the scientific community. It did not take long before he became a member of the National Academy of Sciences of the US. What is striking about Smith is his contributions to science after he received the prize. One might wonder if he should not be invited to visit Stockholm a second time.

The two final acts in the drama of the role of nucleic acids

The opportunities to efficiently and accurately characterize genomes of different forms of life have completely changed our approach to biology. Classification of evolutionary relationships can now be made by sequence analysis of genomes. Each year, information about new important and interesting genomes is being published. Because humans represent the only species that can sequence its own genome the characterization of the double set of three billion bases in the human genome was considered a watershed event. It was possible to deduce that it contains some 23,000 genes that direct the synthesis of different proteins. Similarly the number of protein-coding genes was identified as being about 1,700 in the bacterium *H. influenzae*, about 6,000 in yeast, almost 14,000 in the fruit fly, etc. Humans do not in fact have the most complex genomes among all the species in nature, as we might be prone to think, seeing ourselves as the most "advanced" species. There are many plants, insects and even single-cell eukaryotes, like amoebas, that have much more complicated genomes. The marbled lungfish has a genome that is 40 times larger than ours.

Throughout the 1970s and 1990s there was a focus on the number of polypeptides potentially generated by the linear information in DNA(RNA) and what three-dimensional functional proteins they specified. In the fourth act

Sidney Altman and Thomas Cech, recipients of the Nobel Prize in Chemistry 1989. [From *Les Prix Nobel 1989.*]

of the drama of molecular hegemony proteins had again taken over the scene, but there was something missing in the picture. More in-depth studies revealed that in addition to the linear, predominantly protein-coding genes there was transcription of a large number of additional functional RNA molecules. It was time for another renaissance of nucleic acids in yet one final act.

In 1989 a highly important Nobel Prize in Chemistry was awarded to Sidney Altman and Thomas R. Cech (Table 3). The prize motivation is brief: "for their discovery of catalytic properties of RNA." Sometimes it seems that the shorter the prize motivation, the more important the discovery. Examples are the Nobel Prizes in Physiology or Medicine in 1983 to McClintock "for her discovery of mobile genetic elements" and in 1993 to Richard J. Roberts and Phillip A. Sharp "for their discovery of split genes" (Tables 1 and 2). The latter discovery showed that Beadle's and Tatum's concept "one gene one protein" was not the whole truth. There are situations when a stretch of DNA can direct the synthesis of more than one protein. Maybe the motivation for the prize to Watson, Crick and Wilkins should have been abbreviated to "for their discovery of the molecule of life." What Altman and Cech had found was that in certain cases RNA by itself, in addition to the linear information that it carries, can have enzymatic functions. Thus there are stretches of DNA that code for an RNA that is operative on its own. It is sometimes referred to as *non-protein coding RNA* or — confusingly — non-coding RNA for short. Still the term can be seen as appropriate in the sense that the protein code is not used. The fact that RNA could serve the two functions simultaneously, being an information-carrying molecule and at the same time also an operative molecule, elicited new discussions about the origin of life. Today it is believed that the emergence of life started in an RNA world[55,56] and that double-stranded DNA entered the scene later in evolution as a stable form of information storage which could be replicated with high fidelity. Gerald Joyce at the Scripps Research Institute is one of the leading scientists carrying out research on this dawn of evolution dominated by RNA. On the side he has on some occasions been the master of ceremonies and wine connoisseur of the Thursday Club in La Jolla, where I had my last meeting with Crick.

In 2006 there was one more prize in Physiology or Medicine that further emphasized the role of non-protein coding RNA. Andrew Z. Fire and Craig C. Mello were rewarded "for their discovery of RNA interference — gene silencing by double-stranded RNA" (Table 2). Here was another category of RNA molecules that played a role in the expression of genes. Since then there has been an avalanche of discoveries of various new categories of non-coding RNAs with defined or as yet undefined functions. Expanded and detailed studies of transcripts such as the ENCODE (ENCyclopedia Of DNA Elements) study[57] have revealed that one and the same stretch of DNA can be the origin of a complex mosaic of overlapping, bi-directional transcripts including a plethora of non-protein coding products. The publication describing these important data has more than 200 authors. How would it be possible to single out one or a few Nobel Prize recipients among those? It will take extensive studies before we start to understand the role of all the different forms of RNAs and how they have evolved during the eons of time throughout which life has developed into the complex forms that we see today.

Originally it was believed that only a small part of larger genomes, resulting in protein products — in the human genome about 1% — was of interest. The rest was even referred to as "junk." However, Brenner had already pointed out early on that we should be careful about using such an expression. As he said, junk is something we may save in the attic for unexpected future use, whereas garbage is something that we can dispense with. Progressively we have had to modify our view on the functionality of various parts of, for example, our own genome. It is now appreciated that the major part of the human genome is transcribed into RNA and that the zoo of molecules generated are involved in a highly complex network of regulations.

We are just at the beginning of developing an understanding of these intricate regulation systems. In the end one wonders how this conductor-less network signaling leads to harmonious and functionally interlinked metabolic reactions. Many more insights need to be gained before we understand *homeostasis*, the beautiful balance in an organism through which it maintains a uniform and beneficial physiological stability. The discoveries of all the previously unrecognized forms of RNA have additional implications. The difference between two individuals or groups of individuals and also between us and our nearest relative among the primates, the chimpanzee, is much larger than was originally believed to be the case. Furthermore, in order to understand the influence of changes in genes on development of diseases we need to consider protein-coding as well as non-protein coding genes. Because of the increasing appreciation of the complexity of gene expression, the field of genomic medicine will need time to mature and become practically useful. In total there are many more genes in our genome that do not code for proteins than there are genes that do. It remains to be seen if proteins will attempt another comeback to once again take center stage. Or it may be that discussion about a hegemony of one category of molecule or another is to no avail. Theorell started his laudatory address in 1959[35]:

> *To maa man vaere hvis livet skal lykkes,* "There must be two if life shall succeed", is the theme of a sentimental old Danish song. The author had in mind man and woman, but she probably did not know how right she was from a more elementary biological viewpoint. Two principles are necessary so that "life" shall "succeed." One consists of protein and the other of nucleic acids.

The existence of many different categories of non-coding RNA serving a plethora of different functions has reopened discussions about the definition of a gene. Can a DNA directing the synthesis of an RNA composed of about 20 bases, carrying some critical function, be called a gene? One way of judging this is to evaluate the effects of mutational changes. Regrettably such studies for many reasons do not give a clear yes or no answer. It may seem strange to the reader that, as a result of the expansion of our knowledge, the definitions of the most central concepts of life sciences such as gene and species are subject to discussion. However, this does not mean that the original definitions are called in question, only that they need to be refined and supplemented by some qualitative additions. It is a sign of the health of science when the discussions continue.

Coda — There is still more to molecular genetics

One might think that the growing insight into the molecular details of expressing the information carried by nucleic acids would eventually provide a full picture of the dancing genes. However, there is even more to this complex story. The DNA is wrapped up with the basic histone proteins and packed into chromosome structures. The regulation of the presence or absence of these proteins is of crucial importance. Genes may be activated or silenced depending upon this. One way of influencing this interaction is to add specific chemical groups to either the DNA or the histones. At the present stage it is not known how this is controled. This is just one example of phenomena in the large, more recently emerging field of research called *epigenetics*. This general term is used to signify the emergence of new characters that are *not* due to changes in information-carrying DNA sequences. Epigenetic phenomena are important, for example, in determining that different cells in our body, like liver cells or nerve cells, have distinctive functions. Other kinds of such epigenetic changes are induced by environmental factors. One has registered changes in characteristics of individuals, as for example after extreme starvation that may in fact be carried over from one generation to the next. Such a transfer however requires that the information by some mechanism is assimilated by the germ cells, the only potentially immortal cell lineage in our body. This can happen and therefore, although the absolute majority of mechanisms of evolution are Darwinian, there are special cases when acquired characters can be inherited. Hence it can be said that after all Jean-Baptiste Lamarck was right to a certain minor degree when it comes to developments in Nature.

In another of Koestler's books, *The Midwife Toad and Human Progress,* he discussed the fate of Paul Kammerer. In the early 1900s this Austrian biologist heretically advocated the discredited Lamarckian theory of inheritance based on experiments performed with certain kinds of frogs. He was accused of fraud and shot himself in 1926 in the forest of Schneeberg at the age of 46. Recent studies by evolutionary developmental biologists suggest that the inheritance of acquired traits observed by Kammerer in midwife toads might be real and potentially explained by epigenetic phenomena.

A special case of epigenetics relates to the capacity of proteins to serve as information-carrying molecules. As noted, proteins cannot influence the nucleotide sequence of nucleic acids but they can, in a differentiated way, influence the structure of other homologous proteins. This will be part of the theme of the last chapter.

Chapter 8

Nobel Prizes, Prions and Personalities

In most of the earlier chapters of this book the unique information provided by Nobel archives has been used. This chapter is different, since it concerns two more recent prizes for which no such material is as yet accessible. The understanding of the existence of infectious agents that fundamentally differ from conventional viruses, discussed in Chapter 3, has been recognized by two Nobel Prizes in Physiology or Medicine. The first one in 1976 was awarded to D. Carleton Gajdusek, a prize shared with Baruch S. Blumberg, "for their discoveries concerning new mechanisms for the origin and dissemination of infectious diseases." Blumberg's unanticipated discovery of the nature of hepatitis B virus causing persistent infections was described in the chapter on *Serendipity and Nobel Prizes*. Like Blumberg's finding the discovery by Gajdusek and his collaborators of a completely new form of sub-cellular infectious agents in humans, causing a non-inflammatory brain disease, displayed some of the characteristics of a serendipitous finding.

The second Nobel Prize in the field was awarded in 1997 to Stanley B. Prusiner. This time the prize motivation was "for his discovery of prions — a new biological principle of infection." Originally these agents were referred to as "slow viruses" or simply atypical (unconventional) infectious agents. The term prion was introduced by Prusiner in 1982[1]. It is derived from the initial letters in "*pro*teinaceous and *in*fectious", but in order to make it more pronounceable the position of two vowels of the acronym was swapped. There is a Greek word *prion* meaning "a saw," which has been used for a long time by zoologists as, for example, in the naming of certain kinds of sea birds of the

245

southern hemisphere. However, this did not prevent the introduction of the new acronym form of the word.

Because of my background as a virologist and as an adjunct or full member of the Nobel Committee at the Karolinska Institute for some 20 years, since 1973, the year after I became a tenured professor at the institute, I was heavily involved in the reviewing of Blumberg's and Gajdusek's work. It was my privilege to give the laudation at the prize ceremony. Later on my engagements included reviewing Prusiner's work and the preliminary discussions of possibly awarding a prize to him. However, earlier during the same year when he was to receive his prize I had left the Karolinska Institute to take over the responsibility for the Royal Swedish Academy of Sciences. It was Ralf F. Pettersson, professor of molecular biology, who gave the introductory speech to Prusiner at the ceremony. I have had the opportunity to become personally acquainted with both Gajdusek and Prusiner, two remarkable scientists and personalities. Let me first introduce Carleton Gajdusek.

The encyclopedic eternal traveler

It is almost impossible to give a comprehensive presentation of a person like Carleton Gajdusek. It is a truism that each individual is unique but Gajdusek was exceptional and more unique than most other people. He was a prolific writer and kept an extensive diary during his wide travels, besides his voluminous correspondence. These bound diaries, which are surprisingly personal, were circulated to many of his friends. They are said to cover maybe more than 100,000 pages. I have a collection of many of these diaries in some 20 volumes in my library. A few of the early letters relating to the studies of the disease kuru among the Fore people in New Guinea have been collected in a book with an introduction by Gajdusek[2]. He has also given a very personal description of his life in the volume of *Les Prix Nobel en 1976*[3] that also contains his Nobel lecture. Georg Klein, whom we have already met several times before in this book, has had many contacts with Gajdusek over the years and also owns a collection of Gajdusek's diaries. He has tried, with considerable success, to catch his elusive personality in a long chapter, called "Proteus II — Carleton," in one of his books[4]. In another book, *Deadly Feasts*[5], published about the same time, Gajdusek is the key actor. More recently he appears briefly in the book *Exuberance* by Kay Redfield Jamison[6] and is discussed extensively in the book *The Collectors of Lost Souls*[7]. These different publications will probably not be the last to attempt to understand the uniquely complex and controversial personality of Carleton Gajdusek. Let me give a brief summary of his life until the time when his achievements were recognized by the Nobel Prize in 1976.

Gajdusek grew up in Yonkers, New York. At the time it was a seething ethnic and cultural melting pot. His father, an ebullient and vivacious man from Slovakia, worked as a butcher to support the family with his two sons. His mother, who also had her roots in Eastern Europe, brought different qualities, the more tempered academic and esthetic aspirations. She infused Gajdusek with a deep interest in world literature and folklore. Already before his teenage years he became an avid reader and he retained an unquenchable enthusiasm for literature. Once in 1978 I stayed with my family in Gajdusek's house in Chevy Chase outside Washington. He himself was not at home on this occasion and therefore my wife and I were invited to use his bedroom where he also had his library. Like any personal library it was revealing. It contained most of the best literature created by human civilizations. Reading remained for him an incessant engagement and in 1997, the year in which he was imprisoned, he read among many other things all Moliere's and Shakespeare's plays.

The staircase in Carleton Gajdusek's family house in Yonkers, New York. The names on the stairs are taken from Paul de Kruif's book *Microbe Hunters*. [Photo by Dmitry Goldgaber.]

Gajdusek's interest in science also developed at a remarkably young age. He frequently refers to his aunt Irene. She was an economic entomologist operating in many places around the world. Under her supervision Gajdusek made excursions in the surroundings of their house and to museums of natural sciences in neighboring New York. He seemed to have consumed much of the knowledge that came his way in school or in other contexts and already during his teenage years he got involved in practical laboratory work at the Boyce Thompson Laboratories, located in the neighborhood of his home. It did not take long for young Gajdusek to realize that he should become a scientist. Like a large number of scientists at the time he was inspired by Paul de Kruif's *Microbe Hunters*[8] and he even put the names of the scientists depicted in the book on the steps leading to the attic where he had his chemistry laboratory. However he omitted E. Roux, T. Smith, Ross and B. Grassi and instead added Jenner, J. Lister and Noguchi. One reason Ross was excluded might be that Gajdusek had the British edition of the book. Ross took exception to how he was described and to avoid a libel suit the chapter about malaria was therefore deleted in this edition of the book. The names on the staircase are still there (page 247). Not seen in the picture are stairs higher up carrying the names Reed, Lister, Koch and Pasteur. When one has climbed all the steps one can see, looking back, one more name at the top of the opening of the staircase. That is Hippocrates. The precocious young Gajdusek clearly knew his future mission.

After some training in mathematics, physics and biology at the University of Rochester and an inspiring training in marine embryology under Victor Hamburger at Woods Hole Marine Biology Laboratories, he entered Harvard Medical School. He graduated from there at the age of 22, one of the youngest graduates ever. His nickname at the school was the "burning meteor" which says something about his unusually advanced and intense personality. At this early stage of his career he developed an ambition to seek out the remaining untraveled parts of our world and to attack diseases assumed to be incurable. During his teenage years he had come to develop a passion for mountaineering, hiking, canoeing and camping, equaling his obsession with science.

It is a well documented experience that forthcoming Nobel Laureates have often worked in the laboratory of other (forthcoming) laureates[9]. No one exemplifies this better than Gajdusek. He moved to California Institute of Technology where he interacted with Linus Pauling, Max Delbrück (Chapter 3), George Beadle (Chapter 7) and a wide range of other excellent scientists. Besides the exciting science there were opportunities for a number of outdoor activities with different groups, often with Gajdusek leading the strenuous camping and hiking to deserts and mountains in the western parts of the US, Canada and Mexico. In James D. Watson's book *Avoid Boring People*[10] it is described that "Subsequent non-stop lab orgies, during which I was in the lab long past midnight, alternating with manic weekend car trips instigated by the indefatigable Carleton Gajdusek, who had completed his degree at Harvard Medical School two years before and now was supposedly getting postdoctoral experience in both Max's (Delbrück) lab and the chemist John Kirkwood's." The paragraph finishes "… interrupted our journey onward to Ciudad Obregón, where 110-degree temperatures finally persuaded Carleton that you could die from the heat. On subsequent weekends, Carleton's extreme traveling turned towards the much cooler Sierras, where on one occasion the rest of our party reached the summit of Mt. Whitney long after he had gone on and descended into a valley to the west."

From California Gajdusek returned to the east coast of the US and worked with John F. Enders at Harvard University (Chapter 5). During these postgraduate years he obtained a deep insight into clinical pediatrics and also neurology, working at the Children's Hospital in Boston. His interest in children, their development and their diseases never waned. In 1951 he was drafted to complete his military service with Joseph Smadel at the Walter Reed Army Medical Service Graduate School. Few people came to mean as much to Gajdusek's development as Smadel. He saw the potential of this hyper-energetic and exceedingly talented young scientist. Under Smadel's guidance he started to travel to all the different corners of our world to find the cause of a range of highly enigmatic emerging diseases. He frequently came up with unique solutions. During these years he visited Iran, Afghanistan, Turkey, the Himalayas, Malaysia, New Guinea and several countries in South America.

On Smadel's initiative it was arranged for Gajdusek to be a visiting scientist for some years, starting in 1954, in Australia in the laboratory of another future Nobel Laureate, F. Macfarlane Burnet (Chapter 3). Gajdusek made some fundamental observations in immunology in that laboratory which according to his own remarks[2] put Burnet on the track towards the hypothesis of clonal selection of cells as the basis for acquired immunological tolerance that gave him the Nobel Prize in 1960.

It was when Gajdusek was in Australia that he learned about a unique epidemic among one of the peoples in the highlands of New Guinea. What was planned to be a brief visit to partly unexplored parts of this island turned into a deep life-long engagement with different groups of primitive people, their cultures and diseases in those regions. Gajdusek made strenuous hikes into the highlands of New Guinea and got to know populations that had never been in contact with western civilization before. He lived with these peoples for varying time periods, helping them with their diseases and collecting blood samples. It has often been referred to that in those days he was a crew-cut maverick with an unwashed T-shirt and shorts (page 250). The unique lifestyles of the peoples he encountered were recorded in his extensive diaries and filmed. One of the peoples, the Fore, numbered about 35,000 individuals living in some 160 villages. Among them there occurred a unique and devastating disease called *kuru*, an indigenous word for trembling. The disease to a major extent afflicted

Gajdusek (left) and Vincent Zigas examining a child kuru victim in 1957. [Gajdusek family collection.]

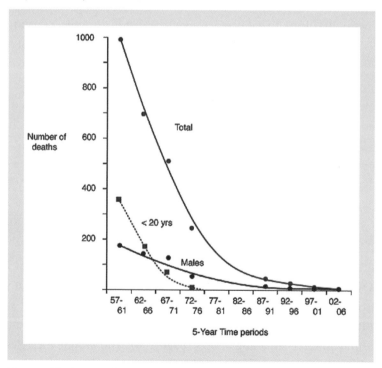

Figure 1. The kuru epidemic. Most cases occurred in women and children, but no child born after 1960 has developed the disease. Cases have occurred into the 21st century. [Figure provided by Paul Brown.]

women and also children, but more rarely men (Figure 1). Together with Vincent Zigas, a district medical officer, he gave the first official description of the disease in 1957[11]. The epidemic had started some decades earlier and cases occurred in increasing numbers. Once the first symptoms had developed there was a steady downhill course with death within 6–12 months. The cause of the relentless disease was a progressive destruction of the brain. There were speculations whether the disease might be genetic, toxic or infectious. The latter alternative seemed less likely since there were no signs of fever or inflammation.

Because of the trust that Gajdusek had built up living with the Fore people he was allowed to make autopsies of kuru cases. In the beginning fixed samples of brains from deceased patients were sent to Burnet's laboratory in Melbourne, but later on, as some early major problems of conflicting interests had been resolved, also to Washington. A first-class neuropathologist Igor Klatzo at the National Institutes of Health (NIH), Bethesda, Maryland, examined the brains of kuru victims. He identified the changes to be similar to those seen in brains from patients with another progressive disease in that organ, Creutzfeldt-Jakob disease (CJD). The pathology was referred to as *spongiform encephalopathy*, meaning that there were "foamy"-like changes in the brain and no signs of inflammation. If changes of the latter kind had been present the condition would have been referred to as encephalitis. The appreciation of the similarity of pathology of brains from patients with kuru and CJD did not help much in the later studies since the origin of the latter disease was not known at the time. However, there was another observation that put the scientists on the right track.

A veterinary pathologist William Hadlow pointed out that the brain lesions of kuru victims appeared very similar to the changes in the brains of sheep with scrapie, a strange infectious disease[12]. At the time Hadlow was a visiting scientist at the British Agricultural Research Council field station in Compton, Berkshire, UK, where he had learnt to know about scrapie from James R. M. Innes, a major UK veterinary pathologist. Hadlow's professional home base was the Rocky Mountain Laboratories in Hamilton, Montana, US, which as we shall see came to play an important role in future prion research.

William Hadlow. [Courtesy of Rocky Mountain Laboratories.]

Milestones in the early studies of atypical infectious agents

The disease scrapie in sheep has been known since the 18th century. It was experimentally proven to be an infectious disease in 1936 by long-term transmittance studies in France. The agent passed through filters that excluded anything larger than viruses. By a curious coincidence it was discovered, while the transmission experiments were being performed, that the scrapie agent had contaminated and survived the formalin treatment of a vaccine against louping-ill — a true viral disease — prepared in sheep. The Icelandic veterinary pathologist Björn Sigurdsson introduced the concept of slow virus infection in the 1950s to describe the disease of scrapie and another disease, visna, also occurring in sheep. Later on the latter illness was found to be caused by a conventional retrovirus, but it took a long time before the nature of the agent causing scrapie was unraveled. The natural route of the spread of scrapie also remained puzzling for a long time.

CJD was identified by two German neuropathologists in the 1920s. The disease is characterized by a progressive destruction of the brain and mostly occur in people of mature age. The frequency of occurrence is one case per million individuals per year all around the world. Certain geographical variations have been identified and a clustering in some families was also recognized early on. The aim of Gajdusek's studies was to make a comparative evaluation of the etiology of scrapie and of CJD. Inspired by Hadlow's observation Gajdusek arranged to have brain material first from kuru patients and then from patients with CJD injected into a wide range of different animals including our closest relative, the chimpanzees. In the mid 1960s it was demonstrated that these human diseases could be replicated in the latter animals and that thus both kuru[13] and CJD[14] were infectious. This was a momentous discovery and it took some careful planning for comprehensive animal experiments to achieve.

In 1958 Gajdusek, supported by his influential sponsor Smadel, had become affiliated with the Institute of Neurological Diseases and Blindness at the NIH. Smadel recommended an appointment at NIH by stating that Gajdusek was "one of the unique individuals in medicine, who combine the intellect of near genius with the adventurous spirit of a privateer." Gajdusek's laboratory was originally labeled Study of Child Growth and Development and Disease Pattern in Primitive Cultures, reflecting his unique and wide-ranging interests. As the engagements developed it came to be renamed Laboratory of Slow, Latent and Temperate Virus Infections. In order to conduct experiments with chimpanzees and other kinds of animals Gajdusek needed new facilities in a secluded environment. After

some negotiations, initiated in 1961 between the
Department of Interior and the National Institutes
of Health, it was decided to choose a facility oper-
ated by the Rare and Diminishing Species Program
of the Interior's Fish and Wildlife Service. It was
the Patuxent Wildlife Research Center in Laurel,
Maryland, which had an isolated location between
Washington and Baltimore. In order to run the
experimental animal operation Gajdusek recruited
one of Smadel's collaborators, Clarence Joe Gibbs,
Jr., who at first was reluctant to leave his studies
of insect-borne infections, but who later became a
very devoted and important collaborator in the team
Gajdusek built up.

Joe Gibbs. [Gajdusek family
collection.]

In the late 1960s I was introduced to him by a colleague in virology at the
NIH. Gajdusek brought me to the Patuxent Center in his Volkswagen Beetle.
It was a memorable experience to be face to face with the chimpanzees that
were used for the experiments. Their striking individual personalities and
close resemblance to us humans made it difficult to see them as experimental
animals, possibly subjected to induced pain and ill-being. However, it was their
subjection to vicarious suffering that allowed the identification of a new kind
of infectious agents in man. Gibbs himself became hesitant about inoculating
these animals with lethal diseases after a few years' experience. Fortunately
it became possible later to transmit the infection to other kinds of animals,
including also some non-primate hosts, as experienced by the use of Noah's
approach — the inoculation of a variety of animals of different species.

When Gajdusek and collaborators had demonstrated that kuru was
infectious it remained to explain the route by which the agent causing the
disease was spread. It took until the late 1960s before it was fully understood
that the situation of transmittance was the ritual cannibalism that the Fore
people had practised for several generations[15]. As an act of mourning, a meal
was prepared from the deceased relative. Women prepared the meal and
came in contact with the different parts of the body including the brain, the
most infectious organ. Apparently children were also exposed, since some
of them contracted the disease. Infection of men was more uncommon. The
introduction of Western civilization caused the Fore people to discontinue
the ritual cannibalism around 1960 and no children born since then have
contracted the disease. However, cases of kuru have still continued to occur

Gajdusek (left) and Paul Brown. [Gajdusek family collection.]

in the Fore people into the present century, showing that incubation times can exceed 40 years[16] (Figure 1).

It was important to understand that the infectious agent causing kuru had been spread by cannibalism, but this left open a number of questions regarding other possible means of the spread of prion diseases. Brains were found to contain by far the greater part of all infectivity, but some infectious activity was also found in other parts of the body, in particular the lymphatic system both in humans with kuru and in sheep with scrapie. Attempts were made to infect chimpanzees and other experimental animals with infectious kuru brain material by the oral route. In most cases no transmittance of disease was seen and it was concluded that a possible spread this way was very inefficient. Attempts were further made, unsuccessfully, to transmit kuru to experimental animals with material from organs other than the brain. It was deduced that in the natural situation CJD is not transmittable. However, in some very special cases of health care procedures it was found that CJD infectivity could be spread between people from one brain to the other. This has been referred to as "high-tech cannibalism" by Rhodes[5] and "friendly fire" by Gajdusek's close collaborator Paul Brown. The professional term used is *iatrogenic,* Greek for physician-borne, spread. It was Gajdusek's laboratory that for many years led the studies into possible ways of iatrogenic spread of CJD.

The first identified case of this kind was the transmission of the disease by a corneal transplant from a patient with CJD. It took 18 months for the disease to develop in the recipient of the transplant. The next cases observed were in a very special situation when an electrode for making electroencephalogram readings by recordings inside the brain tissue was used in consecutive patients. The electrode was first applied to the brain of a patient who was discovered later to have CJD and then to the brains of two younger patients. Because of the sensitivity of the electrode only limited, chemical, sterilization was possible. Both the younger individuals developed CJD within a year. Another neurosurgical procedure, the grafting of cadaveric dura mater, the membrane covering the brain against the skull, was found to be one more means of spreading the disease. To date a total of 196 cases of this kind, mostly in Japan, have been identified[17]. When this means of transmission had been identified the use of dura mater grafts was banned in Great Britain in 1989 and in Japan in 1997. In other countries various precautionary measures were introduced, including selection of donors, introduction of an additional disinfection step and raising the awareness of risks among health personnel working with brain material. Cases of CJD caused by this form of iatrogenic spread now seem to have disappeared (Figure 2, page 265). The two last cases occurred in 2008.

Yet another important source of spread of infection turned out to be growth hormone prepared from pools of cadaveric pituitary glands, a form of transmission to be further discussed below. This latter material is administered peripherally and thus does not allow a direct contact with the brain of the patient being treated. Apparently prions could under special conditions (high dose, access to nerve routes?) spread from the periphery of the body to the brain. Because of this observation the possibility of spread of prions by blood transfusions has been discussed extensively. All the evidence gathered to date does not support this as a means of transmission of spontaneous CJD in humans[18].

Only relatively recently has it been possible to show the transmission of scrapie by blood transfusion[19]. Animals with scrapie have been observed to scrape against, e.g., fences — hence the name of the disease — and this has been deduced to be due to itching as one of the symptoms associated with the disease. Scraping of two animals at the same place could be a source of spread of blood from one to the other. An alternative source of blood spread is by ingestion of placentas left in the field in connection with deliveries of lambs.

As more was learnt about CJD it was found that the overall majority of cases were spontaneous, but in addition there was the clustering of cases in certain rare families. Cases occurring after iatrogenic spread represent a third category of the disease. Prions can be isolated in experimental animals from all these kinds of cases. However, the changes in the brains from patients with the disease may show some differences and such differences can often be reproduced in infected animals. The mechanism of disease, discussed more extensively later, is an aggregation of a particular protein in the body. It is assumed that critical mutations in the gene directing the synthesis of this protein have occurred during the life course of cases, which later develop spontaneous CJD. The increased incidence of CJD in certain families in turn is interpreted to be due to *inherited* genetic variations in the gene that markedly increases the risk for protein aggregation, when an additional critical spontaneous mutation is added.

A unique Nobel lecture and a remarkable extended family

In his Nobel lecture on December 13, 1976 Gajdusek presented the results of the studies he and his colleagues had performed. At this time the Karolinska Institute still strictly followed the formulation in Nobel's will that says that an awardee shall deliver a lecture *after* he has received the prize. Out of courtesy to prize recipients they are nowadays allowed to give this lecture before they have received the prize at the ceremony on December 10. Gajdusek used the title *Unconventional viruses and the origin and disappearance of kuru*[20] for his lecture. I still remember this lecture vividly. It was given in the auditorium of the Karolinska Hospital and I was the chairman. On this occasion the first lecture was given by Baruch Blumberg, who during his allotted 45 minutes told the fascinating story of how hepatitis B virus was discovered (Chapter 2). After him came Gajdusek, who has never been known to stick to a given time. He kept his audience spellbound and before we knew it two hours had passed. The intensity of his presentation combined with a movie he showed made us almost feel like we had physically visited the highlands of New Guinea and been confronted there by cases of kuru. After the lecture the chairman of the Nobel Committee, Börje Uvnäs, came up to me and said: "Now I really understand why we have given the prize to Gajdusek."

Gajdusek's visit to Stockholm was memorable in many ways. It is doubtful if he ever previously had had a tie on and now he was expected not only to dress in tie and suit but also to appear in tails (page 257). He adapted well. More complicated was what to do with the eight adopted boys that he brought along. The first night they stayed at the Grand Hotel arriving with their sleeping bags to the consternation of the establishment (page 257). After

Gajdusek receiving his Nobel Prize from the hands of His Majesty King Carl XVI Gustaf. [© Scanpix Sweden AB.]

Gajdusek and foster boys prepared for the Nobel week. [Gajdusek family collection.]

this they moved to the home of one of Gajdusek's good friends in Sweden. The virologist Arne Svedmyr, his wife Birgitta and their five sons provided an inviting environment for the boys during their week of unique experiences in Stockholm.

Throughout the years Gajdusek adopted a large number of children, the overall majority of them boys. They were of Micronesian and Melanesian, mostly New Guinea, origin. It can be questioned if the term adopted is appropriate or if they should be referred to as foster children. Most of the children were considered adopted in their traditional culture. It was a dramatic cultural change for the 8- to 12-year-old children to move from their original indigenous culture into the modern civilization of the US. Most of them had known Gajdusek since their early life. The total number of children in whose education Gajdusek became involved is estimated to be between 40 and over 60. The reason the figures vary is that, in addition to bringing more than 40 of them into his household, Gajdusek also supported the education of a large number of children in their home countries. It was Gajdusek's ambition that the children brought to the US should melt into their new environment and get an education like other American children. This was achieved only to a variable degree. It should be recollected that the foster children were late in becoming literate and that they originally came from much less competitive societies than the one they encountered in the US. A few of the children made it through two to four years' university education, whereas some of the others attended vocational schools instead. Most of them returned home and in many cases they eventually fulfilled important functions in their country of origin. A few stayed in the US. By these unique efforts Gajdusek came to have a very important influence on the lives of an impressively large number of young children. They called him "father," but only two of them hyphenated Gajdusek with their last name. There were many opportunities for exchange of affections in the large family.

Gajdusek's extended family meant the running of a large household. As in any large family there were a number of practical daily things to manage. One may question how this was managed since Gajdusek was traveling most of the time. And when he was home he expanded the program for his children. He wanted them to tour museums and art galleries and to participate in various cultural events. All these activities took a lot of resources. He estimated that he could run his extended family for two years by use of the money he received in association with his Nobel Prize. Somehow he managed to keep a financially sound balance running his large family throughout many years.

For a long time the large household resided in Chevy Chase outside Washington, but Gajdusek also retained his original family home in Yonkers. He regularly took the children there to experience the culture of nearby New York and to pay tribute to Herman Melville's grave which Gajdusek had discovered in a neglected state across the city line from Yonkers, in the Bronx. When he was not home Joseph H. Wegstein, a computer scientist and interesting person in his own right, managed the large household. This arrangement lasted until 1979, when Wegstein moved out of the house for a brief marriage. Around 1982 the big Gajdusek household moved to a house in Prospect Hill in Frederick, Maryland. A year later Wegstein was again united with the large family, but now sadly he had developed a severe case of Alzheimer's disease. He was cared for by the kids and a home help. Many scientists who were his colleagues can testify to the unique atmosphere at the Gajdusek homes, with dinners almost always including visiting colleagues or other particular friends. Georg Klein had written about this[4] and the Svedmyrs, who took care of the boys during the Nobel week in 1976, had many contacts with the Gajdusek family over the years. They stayed for three years in an apartment in Frederick near the Gajdusek house during the 1990s, allowing many close interactions.

As already mentioned my family and I, including our three children, at the time 9–13 years old, stayed in Gajdusek's house outside Washington in 1978. In Gajdusek's absence, our kindly host was Wegstein. The household at the time included some 8–10 adopted children. They all had their different allotted tasks. Keeping their rooms in order, doing their homework, caring for visiting guests, arranging for the dinner by assisting in the cooking or laying the table. In between they interacted with our children. Basketball was played in the courtyard and lying on the floor they pointed out in atlases from which part of the world they came. Considering the large household everything was in surprisingly good order thanks to Wegstein's discreet management. It was a unique and memorable experience to temporarily be a part of the Gajdusek household.

A San Francisco neurologist challenged by a case of CJD

In his Nobel lecture Gajdusek referred to the remarkable properties of what he called an unconventional virus. No virus-like particles had been observed by electron microscopy in surveillance of many samples from diseased humans and animals. Furthermore the infectious activity of the causative agent showed a remarkable resistance to physical and chemical treatments. These characteristics had made scientists speculate about the possibility that it might replicate without participation of nucleic acids[21,22]. Many scientists considered this heresy, in particular since the existence of strains of scrapie with distinctly different properties had been identified. However, in the late 1970s interesting things started to happen in a laboratory in San Francisco.

Before our family's visit to Gajdusek's house in 1978 we had spent five months in California, where I had studied virus infections in the brains of experimental animals in Kenneth Johnson's laboratory at the Veterans Administration Hospital in San Francisco. During the early part of this visit I had an opportunity to participate in a virology congress in Keystone, Colorado. During the bus ride from Denver airport to Keystone I happened to sit beside Prusiner, who was then in an early phase developing his work on scrapie in San Francisco. We had a very intense and interesting conversation. He described to me his background and how he had become engaged in studies of scrapie as a model for unconventional infectious agents. This contact became the start of a long-lasting friendship that came to give me a unique opportunity to closely follow Prusiner's forthcoming frontier-breaking pursuit of science.

Like all other Nobel Laureates Prusiner has provided a brief history of his life. From this it can be extrapolated that his grandfather was a Russian Jewish immigrant who came from Moscow to Sioux City, Iowa. The family originally came from the village of Prushany, the source of the family name, in present-day Belorussia. Prusiner is yet another example of a Nobel Laureate of Jewish ethnic background who has his roots in the eastern part of Europe (Chapters 1 and 7). Prusiner studied at the University of Pennsylvania where he majored in chemistry. He enjoyed the intellectual environment at this university very much. Many of the teachers were of a high standard and besides chemistry he enjoyed undergraduate courses in philosophy, architectural history and Russian history. His chemistry studies led to his involvement in a research project on hypothermia. He initiated work on brown adipose tissue, which is crucial for temperature regulation, under the supervision of the famous chemist Britton Chance. An extension of these studies led to a year as a visiting scientist at the Wenner-Gren Institute in Stockholm before Prusiner returned to Philadelphia to finish his medical studies. Further training in scientific work was gained in Earl Stadtman's laboratory at NIH.

In 1972 Prusiner started a residency in neurology at the University of California in San Francisco. A case of CJD that he experienced stirred his interest in atypical infectious agents and he initiated some early studies of scrapie as a model system. Progress was very slow but the situation improved after Prusiner had established collaboration with William Hadlow, the same scientist who inspired Gajdusek, and Carl Eklund at the Rocky Mountain Laboratories. This laboratory was managed by NIH, but in 1978 it was decided to shut down the work on scrapie. This forced Prusiner to build up his own resources for scrapie work in San Francisco. In 1980 he visited New Guinea to see and examine victims of kuru (right picture).

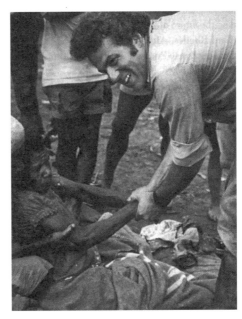

Prusiner examining a kuru victim during a visit to New Guinea in 1980. [Gajdusek family collection.]

The unconventional virus is not a virus after all

Prusiner's aim was to purify the scrapie agent and to define its chemical nature. The ways in which he approached this are described in his Nobel lecture[23]. The early work developed very slowly. Each experiment took more than a year. The first challenge therefore was to find a faster assay to measure the concentration of the agent. The scrapie agent could conveniently be propagated in mice and hamsters, but the reading of the endpoint of infectivity by scoring dead animals was time-consuming. Prusiner, inspired by an approach previously used by virologists, developed a modified assay and used the average incubation time as a measure of amount of agent present in the inoculum. This allowed tests to be finished in 70 instead of 360 days. Using this faster assay he could accelerate his experiments.

After having tried a number of separation techniques and treatments with different enzymes he found that the main part of the infectivity was associated with a protein fraction with a molecular weight of 27–30 kD[24]. At this stage Prusiner had become more and more convinced that the infectious capacity was associated with protein and not nucleic acid. He therefore boldly, and to the chagrin of a large part of the scientific community, introduced the term prion[1]. The protein he had found was referred to as *prion protein* PrP. In this early phase he was frequently exposed to severe criticism for the heretical data that he produced. He reacted promptly to this and he has therefore been referred to as "genial, prickly and aggressive."[7] On occasions it became my mission to encourage Prusiner not to become too defensive and immediately go into counter-attack. The advice I gave was to acquire a more diplomatic hesitance and to let the data speak for itself.

In the spring of 1984 Prusiner and I participated in a meeting in Scotland where he was the main speaker and I was invited to summarize the proceedings. Most of the other participants were from Alan Dickinson's group in Edinburgh, which had long been doing extensive work in the scrapie field. They were highly critical of Prusiner's work. There were a lot of discussions about prion strains, the pathogenesis of the disease and the nature of the agent. It very soon became clear that the purpose of the meeting was to demolish the "protein only" concept developed by Prusiner. How could an agent containing exclusively protein cause diseases with markedly different incubation times and highly divergent pathology? Prusiner defended his case well but the atmosphere was close to hostile. After the meeting the two of us had half a day to spend in Edinburgh before flying home. There was a lot to talk about when walking the streets and enjoying the cafés of this historic city. Maybe this home of the famous Scottish Enlightenment and the illustrious Moonlight Society at the end of the 18th century[25] provided the proper sounding board for our exchange of ideas. The charm of science is that it sometimes generates very intense intellectual discourses and that the sharing of these may lead to a rich friendship. Just at that time Prusiner and his colleagues were on their way to making a number of crucial breakthroughs in their research.

It was discovered that Prusiner's PrP material was pure enough — it contained predominantly a single protein — to identify a short amino acid sequence in one of its ends. This in turn made it possible, by use of molecular cloning, to identify the gene responsible for the production of this protein[26]. It was then found that it was a *normal* host cell gene that coded for the (dominating) component of the infectious particle. All of a sudden this finding made the infectious process comprehensible. It was possible to understand why there was no inflammation and no immune responses in connection with prion diseases. No such reactions would be expected to be seen against a normal protein in the body. This is due to the phenomenon of immunological tolerance, recognized by the award of a prize to Burnet and Medawar in 1960 (Chapters 3 and 6).

The next important step in the development of the studies was to analyze what would happen if one eliminated the gene responsible for the PrP protein. This is possible by use of a so-called "knock-out" technique, an advance in biomedical sciences recognized by a Nobel Prize in Physiology or Medicine in 2007 to Mario R. Capecchi, Martin J. Evans and Oliver Smithies "for their discoveries of principles for introducing specific gene modifications in mice by the use of embryonic stem cells." An expression of the PrP gene is observable in cells already during the early stages of embryonic development and the protein might therefore have a fundamental function. Thus it was uncertain if animals could survive without a functioning PrP gene. To the delight of the scientists the animals lacking the gene not only survived, but seemed to function normally and had a normal life span[27].

Access to mice without a functioning PrP gene allowed a very critical experiment. What would happen if one infected such animals with high doses of the scrapie agent? The result of this experiment was stunning. They were completely refractory to the infection[28, 29]. In addition these animals could develop antibodies against PrP, since they were no longer tolerant of the protein. Animals lacking the PrP gene were also found to be useful for many other purposes. PrP genes from other species could be inserted instead of the deleted original mouse gene.

Prion proteins from different species vary more or less in their structure. These differences are important since they are the reason why prions from one species often cannot infect another species. There is a *species barrier* to infection. Thus, for example, one has never seen an infection with scrapie prions in man. This barrier can be overcome by use of transgenic mice in which the normal PrP gene has been replaced by the corresponding gene from another species. Insertion of the PrP gene from humans has allowed studies in mice of different forms of homologous genes. For example the importance of mutations identified in families with an increased frequency of CJD can be studied experimentally. More than 40 different mutations capable of influencing the reactivity of the PrP protein of man have been identified. By the transgenic approach, mice also can be made sensitive to prions from bovines, something that turned out be practically useful as we shall see.

Prusiner and his colleagues as well as other research groups have also used transgenic mice for clever studies of the nature of strain variations of prions[30, 31]. In complicated experiments they infected animals carrying different PrP gene constructs with different levels of expression with regular scrapie infectious material or different incomplete PrP gene constructs. The infected animals developed disease with different incubation times and varying kinds of changes in the brain. Thus it appears that one and the same normal PrP protein can be induced into different foldings under the influence of different partly homologous pathogenic PrPs.

An alarming epidemic of CJD in younger patients

On April 23, 1985, I was called out of a meeting with the Swedish Medical Research Council for a consultation with the Kabi Company in Stockholm. For some years this company had been one of the main producers of growth hormone, an important remedy for children showing slower than normal growth. The hormone was prepared from pools of thousands of pituitary glands collected indiscriminately in connection with regular autopsies of humans. Thus brain material from patients with CJD might well become included. At the meeting I was told that four cases of CJD had been observed in the US in patients treated with growth hormone preparations produced by this procedure by other pharmaceutical companies. What to do? My advice was to immediately stop all production of pituitary-gland-derived growth hormone and to recall all distributed products to prevent it from being used. After an extensive discussion this was accepted by the company.

This decision caused some major financial losses, both because of loss of sales and also because compensation had to be paid to contracted suppliers of pituitary glands. However, the saving grace of the situation was that the company was in the forefront of development of a production of recombinant DNA generated growth hormone in bacteria. Efforts to produce this kind of drug were accelerated and eventually the Kabi Company was the first one to introduce the new and safe form of biosynthetic growth hormone for substitution therapy. The company was also fortunate to note during the ensuing decades that no case of CJD developed in any patient treated with pituitary-gland-extracted growth hormone that it had produced. However, sadly, all the more cases were seen in children treated with this type of product prepared by other companies. The total number of cases to date is 206 and the epidemic appears to have just tapered off[32] (Figure 2). Most cases have occurred in France — 109 people out of 1,700 people treated. The corresponding figure in Great Britain is 56 out of 1,848 and in the US 28 out of 7,700. In the latter country an additional purification step was introduced in 1977, which seems to have greatly reduced or even removed the risk for transmission of the infectious agent.

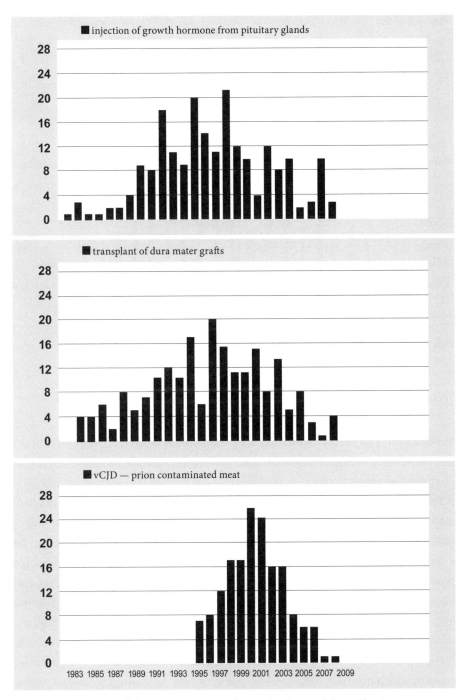

Figure 2. Comparison of three unique CJD epidemics. Each one of the epidemics encompasses about 200 patients and they were all caused by different human interventions; iatrogenic CJD by intra-muscular injection of pituitary gland-derived growth hormone, iatrogenic CJD by dura mater transplants and vCJD by ingestion of contaminated meat. The average incubation time of the three epidemics was 15 , 11 and most likely 11–12 years, respectively.

Nobel Prizes, Prions and Personalities 265

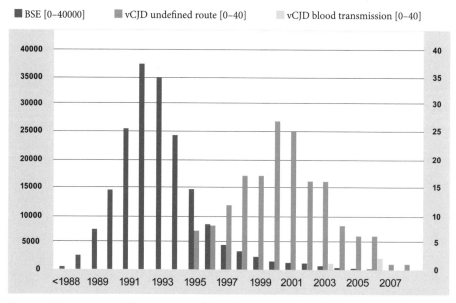

■ BSE [0–40000]　　■ vCJD undefined route [0–40]　　■ vCJD blood transmission [0–40]

Figure 3. The epizootic of bovine spongiform encephalopathy — "mad cow disease" — (blue columns), the epidemic of vCJD in humans (red columns) and three cases of transmission of vCJD between humans by blood transfusion (yellow columns).

"Neo-cannibalism" in cattle and a new form of CJD in humans

Towards the end of his Nobel lecture[23] Prusiner mentioned the recent prion disease among cattle in Great Britain. "Mad cow" disease or more appropriately bovine spongiform encephalopathy (BSE) epizootic started in 1985/86. It reached its peak in 1992, but still into the present time occasional cases of infected animals are detected (Figure 3). More than 180,000 cases of infection so far have been identified. The source of the epizootic turned out to be a new kind of feed mainly for dairy cows, which have a particular need for protein. Their natural feed of plant proteins was supplemented with meat and bone meal. The latter was prepared from the offal (the *off-fall* of slaughtering and butchering) of sheep, cattle, pigs and chickens. This mixed food had been used for some time but in the late 1970s a previously used hydrocarbon-solvent extraction method was abandoned and this led to a markedly increased fat content of the product. This modified kind of feed was found to spread prions of either bovine or sheep origin. When this was recognized in 1988 its use was forbidden. One year later slaughtering techniques were altered to prevent brain and spinal cord from being present in products for human consumption. This second measure was taken to provide an extra assurance that prions would not spread to humans. At the time it was not expected that diseases caused by such a spread would be seen. It was inferred that a species barrier, which through hundreds of years had been experienced to prevent the spread of sheep scrapie to man, would also exist between bovines and man. One more effort to further increase the safety was extensive culling of cattle on farms where an infected animal had been identified.

In spite of all the measures taken it was learnt that prions from animals with BSE could spread via food to humans. The first case was identified in 1994. It was proven by molecular methods that the disease was caused by a new variant of prions infecting humans and it was therefore referred to as vCJD[33]. This form of CJD differed in many ways from the common form of CJD and added a fourth category of human CJD to the three described above. The average age of the patients was much lower, 26 years, and the histopathological changes in the brain had different characteristics than the most common form of CJD. This new form was possible to mimic in transgenic mice carrying the bovine PrP gene by infecting them with vCJD prions from humans.

The observation that BSE could spread to man led to further culling of animals. In 1996 all British cattle older than 30 months were slaughtered. During the dawn of this epidemic there was a great deal of speculation about how extensive it would become. Fortunately it turned out that the epidemic peaked in 2000 and that in 2010 it is coming to an end (Figure 3). The total number of humans afflicted has been 214, out of which 167 cases have occurred in Great Britain, 25 in France and the remaining cases are scattered, mostly in European countries[34]. This epidemic added a third one to the above-mentioned two epidemics caused by iatrogenic spread of prions via pituitary-gland-extracted human growth hormone and grafting of dura mater. All these epidemics, afflicting similar numbers of people (Figure 2), were caused by human interventions, although of very different kinds. Disease developed in patients receiving subcutaneous injections of growth hormone at an average time of 15 years (range 4 to 36 years) after the midterm of their treatment; in patients receiving the dura mater grafts, allowing a direct brain surface to brain surface contact, at an average time of 11 years later (range 16 months to 23 years); and in cases of vCJD, most likely infected by eating contaminated meat, after about 9 years judging from the interval between the peaks of the BSE epizootic and of the vCJD epidemic (Figure 3). However, it should be noted that the apparent incubation time in the latter case is most probably longer, since presumably no — or only very little — contaminated meat had been distributed after 1989, when the slaughtering technique was altered to exclude the presence of brain and spinal cord material in the final food product.

Neo-cannibalism in humans?

Nature never wastes any material. Organisms, ranging from the simplest to the most complex, represent rich sources of complex molecules and energy-rich material to be recovered after death. One rarely sees carcasses in Nature, except perhaps some bone remains. All organic material is rapidly processed by various scavengers, not the least microorganisms. Theoretically, consuming your own species, *cannibalism*, could be of value, but it could also increase the risk of spreading infectious agents, which have adapted to use a particular species as their host. There are infections that have very limited potential to spread even within the selected host species. One example is the hepatitis B virus infections described in Chapter 2. Chronic infection by the virus can only spread by blood and under natural conditions transmission of blood from one person to the other is rarely seen. However, there is one exceptional situation. This is pregnancy, which results in an exchange of cells from one generation to the next — the pregnant woman, to the offspring — both during gestation and at delivery. Hepatitis B virus has survived in Nature by such a blood spread, a so-called *vertical infection*, through history.

In spite of the risk for spread of particular infections, cannibalism is not uncommon in nature. However, it seems that it is mostly practised in particular situations where it may have significance for evolutionary developments. There are many examples. Among the at least 100 species of mammals that practice cannibalism one finds females that eat stillborn or deformed babies, and males, as in the case of lions and chimpanzees, that kill and sometimes eat cubs and baby chimpanzees to force their mother into oestrus, so that they can sire the next litter. In us humans the practice of cannibalism is strongly influenced by our accentuated recognition of pronounced individual personalities. Three forms of cannibalism have been identified in human history. One is ritual cannibalism, such as that practised by the Fore people as an act of mourning. In this case the practice came to illustrate the risk of spreading an endogenous infectious agent. Another is "victory" cannibalism, in which case a conqueror humiliates and extracts the strength from the killed, defeated individual(s) by eating them. Finally there is survival cannibalism when the only possibility for members of a group to continue living is to eat other dead or killed fellow humans. This may occur only in very rare and particular extreme situations, as in historical events when a lifeboat carrying survivors has been out at sea for a long time or more recently when a fraction of the passengers managed to survive a plane crash in a very isolated part of a jungle.

Although cultures in the modern global society essentially exclude cannibalism there are an increasing number of situations in modern medicine when tissues or organs from one individual are transferred to another human being, the previously-mentioned high-tech cannibalism. Blood transfusions have been used ever since the fundamentals of the human blood groups were discovered (Chapter 5) and during more recent decades transplantations of bone marrow and solid organs have become common practice in human health care. This became possible when the nature of the immunological differences between individuals was recognized and as a consequence the barriers for transplantation could be reduced (Chapter 6). The fact that patients receiving transplants have to be immuno-suppressed in order to prevent rejection of the foreign tissue further complicates the situation. One downside of this suppression is that it may allow replication of viruses possibly carried over by the graft and also dormant viruses in the recipient of the transplant. Most often these activated infections can be managed. However, this problem of activation of a silent infection differs when it comes to prions, since they are not controlled by the immune defences of the body.

When it had been identified that vCJD caused an epidemic in humans the question arose whether the agent causing the disease might be spread via medical interventions. As discussed above an iatrogenic spread of spontaneous CJD was only seen under very special conditions. Since one had never seen any spread of the spontaneous form of CJD by blood transfusions it was assumed that vCJD also would not be disseminated by this procedure. It therefore came as a surprise when it was documented in 2004 that vCJD could in fact be spread by blood transfusion[35] (Figure 4). Four cases of infection caused by such a transmission of prions have been documented. In all cases blood was taken from individuals who some years later developed vCJD. It took 5–8 years until the recipient of the contaminated blood developed the disease, which occurred in three of the four cases. In the fourth case the patient died of another disease and the presence of vCJD prions was detected in a tissue sample from the body at autopsy.

Already before these cases of iatrogenic transmittance of prions causing vCJD had been documented, measures had been taken to reduce the risk of their possible spread by blood. Intense attempts were made to develop a

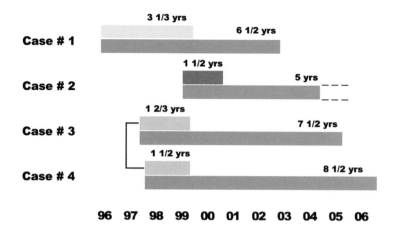

Figure 4. Four cases of vCJD infections caused by blood transfusion. Cases 3 and 4 were infected by blood from the same donor. In each case the upper bar shows the time until the donor developed disease and the lower bar the time until disease appeared in the recipient or as in case 2 vCJD prions were demonstrated in tissues. [Figure provided by Paul Brown.]

blood test which could demonstrate, as in the case of hepatitis B and human immunodeficiency virus infections, whether a potential donor was infected. Regrettably attempts to find a reliable test have failed to date. However, other precautions had been taken. It was known from studies of mice and hamsters infected with scrapie prions that these might occur outside the nervous system, preferentially in white blood cells. It had therefore been decided, first in Great Britain, already in the late 1990s, and then in other European countries, that white blood cells should be removed from blood to be used for transfusion. From animal experiments it was deduced that this would reduce, but not eliminate, the amount of possibly contaminating prions. Since the introduction of this precautionary measure no more cases of vCJD caused by blood transfusion have been seen. As a consequence of the fact that the number of cases of vCJD have progressively decreased since the year 2000, the risks for blood transmissions between humans should diminish progressively.

The mechanisms of prion diseases have a wider application

The identification of the crucial importance of a normal host protein in the development of prion diseases led to two follow-up questions to answer. The first of these concerned the role of the protein under normal healthy conditions,

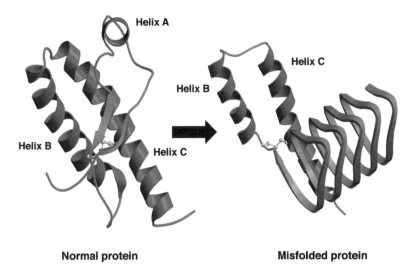

Helix A

Helix C

Helix B

Helix B

Helix C

Normal protein   **Misfolded protein**

Figure 5. The fundamentally different structure of normal and inappropriately folded PrP protein. The latter has a dominating occurrence of beta-pleated sheets, which gives it a propensity to aggregate with other homologous proteins, potentially causing destruction of tissues. [Figure provided by Paul Brown.]

and the other, by what mechanisms events involving the protein could lead to disease. Let us first consider the latter issue. It appears that the disease process is caused by an inappropriate folding of the protein already alluded to. The "healthy" normal protein has a structure dominated by three so-called alpha helix structures, whereas the disease-causing protein has a dramatically different structure[36,37]. Besides two minor alpha helix structures the folding of the latter is dominated by four beta-pleated structures (Figure 5). The paradox of this situation is that it is *one and the same* protein chain that can be folded in these two completely different ways.

The exact mechanism by which the disease-causing PrP can induce a pathological process is not known as yet. It has been inferred that this protein, when in contact with the normal healthy protein, causes the latter to misfold and that this effect can take the form of a cascade, like an "avalanche" phenomenon. It is probable that this spread of misfolding of PrP requires additional components, most likely other proteins. Certain kinds of misfolded proteins have a propensity to form aggregates. Such aggregates are called *amyloid,* a very special clumping of proteins known to pathologists for a long time. It came to be called amyloid, because it was believed, incorrectly, that it represented starch, latin *amylum*. Different kinds of amyloid have been found both in the brain and also outside this organ associated with a number of diseases. The protein aggregates show very specific staining characteristics. Treated with Congo red they display an apple-green fluorescence when viewed by polarized light.

Besides CJD one has identified a number of diseases that also appear to be caused by a similar type of aggregation of proteins, but of different kinds. Examples are Alzheimer's, Parkinson's and Huntington's disease in the brain as well as type 2 diabetes and immunological amyloidosis as well as more general diseases[38–40]. The fact that prion diseases are infectious makes them attractive to study in model systems as prototypes for amyloid diseases. One can anticipate that future studies of amyloid will give important new insights into the mechanism of a number of human diseases and hopefully also include new critical discoveries.

In spite of extensive studies it has not as yet been possible to come to an agreement on the physiological role(s) of the normal PrP. Several different possible functions have been proposed for this generally membrane-bound protein. Among these can be mentioned importance for different signal pathways critical for the survival of cells, protection against oxidative stress, binding of copper and maintenance of the isolation material, myelin, in peripheral nerves. It has also been suggested that it may be involved in memory functions in the brain.

Comprehensive studies have been made of prion and prion-like proteins in yeast. It turns out that these proteins are of importance for a kind of inheritance that is not based on the functions of nucleic acids, so called epigenetics (previous chapter). This non-Mendelian inheritance has been shown to be critical for the survival of yeast cells under certain conditions of stress[41,42]. In this system the underlying mechanism is information exchange between proteins of the kind represented by prions. We are only beginning to understand how this information exchange, outside the central dogma involving nucleic acid information flow (Chapter 7), functions. The diversity of information exchange is exemplified by the capacity shown by certain varying kinds of prion proteins to influence the normal homologous host protein in specific ways so that diseases with different incubation times and with diverging histopathology can develop. Characterization of mechanisms for specification of protein folding represents a challenging field for future studies.

The reason why the field of prion research was chosen for the concluding chapter in this book, in spite of the lack of access of Nobel archival material to review, is that it provides unique examples of the sometimes remarkable and unpredictable advances of new knowledge. The anti-dogmatic intellectual journey from examination of ritual cannibalism as a cause of disease to pathogenetic events in the brain caused by misfolding of an endogenous protein and further on to an emerging insight into a wide range of non-infectious protein misfolding diseases and the existence of non-nucleic acid based inheritance is remarkable. Clearly there are more discoveries to come and potentially more prizes in Physiology or Medicine to be awarded in this subfield.

The archetype of the successful scientist

Let me conclude this book with some thoughts on the personalities of leading scientists. Much has been written about the lives of successful scientists, both by themselves and by others. It is a truism to note that each story is unique, since the personal qualifications, reflecting the intertwined inherited and environmental influences, brought into the process of major discoveries vary with each individual. In several of the previous chapters I have reflected on conditions, both concerning individuals and environments, that further creativity and allow the emergence of major discoveries. One thing is clear and that is that such discoveries cannot be planned for. If life were that simple it would probably be quite boring.

Attempts to understand the enigmatic developments of human creativity, be it in the arts or in science, that further the advance of our global civilization, are an endless source of reflection. Many adjectives have been used in the earlier chapters: curiosity, enthusiasm, adventurousness, heterodox and heretical thinking, etc. In science and technology there is a clear progressive growth of our knowledge and insights into the functions of Nature and the universe. Whether the same progression can be seen in humanistic sciences is debatable. In our attempt to contribute to the endless progression of knowledge we clearly depend on those who in earlier times have contributed to this enterprise and frequently we seek inspiration from insightful, often mesmerizing mentors. As in the case of Gajdusek, contacts with previous or forthcoming Nobel Laureates have often provided guidance in the development of a successful scientist.

The reward to a scientist can be very large if — and this applies to only a few "anointed" (lucky) researchers — they manage to make a fundamental discovery. The Eureka feeling of being the first in our human civilization to have a fundamentally new insight can probably not be fully described. It needs to be experienced. Still, the opening of one previously closed door in scientific endeavors frequently leads to an understanding that there are new, until then unknown, doors to be opened in the future. The discussions about a possible end of science are in my opinion nonsense. There will be new discoveries to be made for a limitless number of coming generations of scientists. This obviously bodes well for the Nobel Prizes of the future. It is clear that the success story of these prizes, evolving over more than hundred years, will continue into centuries to come. But can anything of a more general nature be learnt from studying the scientists that have made major discoveries, which have been recognized by Nobel Prizes? Running the risk of making over-generalizations I will conclude by discussing a few major critical features.

I have had the privilege of meeting a large number of the scientists who have been prime movers in the advancement of life sciences. Their impact, not least after the emergence of molecular biology, has been enormous and it is obvious that we are now set on a course that will lead to amazing and unexpected discoveries in this field during the coming decades, as indicated in Chapter 7. The personalities I have met naturally have widely diverging personal features, but there are certain traits they have in common. They clearly are endowed with a unique intellectual capacity, but there is an absolute need for additional qualities to productively exploit this resource. Jim Watson started his famous book *The Double Helix* — about the discovery of DNA — with the sentence "I have never seen Francis Crick in a modest mood." Successful scientists have an intensity in their personality; they are obsessed by their research endeavors. They therefore demand room for their own personality in professional and also social interactions. Humility is not for them. They act as if they know that they have a mission and their life is centered on their own person. They may not be role models for how life in general should be conducted. This has been documented in many contexts, e.g. in the book *Intellectuals* by Paul Johnson[43]. Focusing on selected examples of certain leading intellectuals through the centuries he reviewed the moral and judgmental credentials they brought to the conduct of human affairs. He found these credentials poor and noted the exposed situation of being an offspring of these influential personalities. They of course dominate, for good and for bad, not only their professional colleagues, but also their family and wide social circles. Their self-centered behavior becomes apparent in many ways.

I once heard a story from a good friend, a scientist from Southern California. The story concerned two of the dominant leaders in modern biology who had been out for dinner. After the dinner one of them called my friend and said, "I can't stand this scientist X. He just talks and talks and I can't get a word in there." It did not take long before scientist X also called my friend and his message was essentially the same, except that he could not stand scientist Y because of his dominating the conversation. A few months later I was at a dinner, including spouses, with my friend and the two scientists X and Y and also the Nobel Laureate Manfred Eigen. It became clear during the dinner that Y was the dominating scientist. Y by the way was Sydney Brenner who, to the great satisfaction of the scientific community, eventually received a Nobel Prize in Physiology or Medicine in 2002 (Chapter 7).

This description of leading scientists may give the impression that they are monolithically engaged in their work. This need not necessarily be the case. These richly-endowed personalities often in parallel can develop other impressive talents. Some of them are excellent musicians. One good example is the above-mentioned Eigen who received his prize in Chemistry when he was only 40 years old. When he celebrated his 70th birthday in La Jolla in 1997 we as guests received a unique complimentary gift. It was a special edition of two recordings that Eigen had made of Mozart's piano concertos in A major, K. 414, and in C major, K. 415. The former was recorded with Eigen playing a hundred-year-old Steinway grand piano accompanied by the New Orchestra of Boston led by David Epstein.

The most incessant talker among scientists I ever met was Gajdusek. In Rhodes' book[5] he was described as "A compulsive talker who spills ideas non-stop for hours — good talk, often brilliant talk and consummate story-telling, but more than some listeners can bear." Similarly Redfield Jamison in her book on exuberance[6] cited the themes jotted down by her husband and herself after a dinner conversation with Gajdusek, as follows: "The very partial list included immunology, love, the French, the Americans, the Dutch, suicide, Puritanism and sex, schizophrenia, rat poison, molecular biology, tuberculosis, the FBI, the idiocy of American politics, Melville, Russian explorers, Plato, anthropology, mad cow disease, New Guinea boys, courage, moods, navigation, linguistics, and meadow mice." At the dining table in our house we have on occasions been exposed to the same firework of diverse subjects, mostly executed in a monologue form by Gajdusek. His profuse talking was difficult to stop. Klein mentioned in one of his books[4] that on one occasion he managed to make Gajdusek listen. This was when he presented Hungarian poems in the original language. Gajdusek, who was fluent in an impressive number of different languages and argued that anyone can learn to manage a new language in five months, in particular at a young age, apparently for once was put in a listening mode by the combination of poetry and foreign language.

Gajdusek's general overwhelming dominance of conversations could even be described as a constant mania. In fact MacFarlane Burnet (Chapter 3) once

said: "Gajdusek is quite manically energetic when his enthusiasm is aroused."
This is referred to by Redfield Jamison in her writing about exuberance and she should know about swings of mood. In other books she has written about the relationship between creativity and psychotic diseases, the fuzzy border between genius and madness. In *Touched with Fire*[44] she has reviewed relationships between manic-depressive illness, now often referred to as bipolar disorder, and the artistic temperament. It is quite clear that sometimes the markedly dominating behavior of leading scientists comes close to the relative border of what we call normality. To what extent the molding of their personalities is a product of the environment or genetic conditions they carry along is a matter of debate. The offspring of successful scientists may therefore carry a double cross, having genes that may accentuate the swings of their personality and growing up in an environment dominated by the scientific pursuits of a parent. Still it is important not to over-generalize. There is in fact considerable evidence for the opposite. The homes of (future) Nobel Laureates can provide an excellent environment to bring up Nobel Prize recipients in the next generation. In the previous chapter it was related how two of Arthur Kornberg's sons came to be very successful scientists, one of whom even received a Nobel Prize, and, as was also pointed out, there are six more examples of multi-generational Nobel Prizes.

My last encounter with Gajdusek was in November 2006. He was in Stockholm to lecture in a series on *What is life?* arranged by Ingemar Ernberg, professor of tumor biology, at the Karolinska Institute. The title of the lecture was "What is life? What we learned from cultures which had never before met civilization — origin of man." Gajdusek gave a broad exposition of his rich experience of original indigenous cultures and showed a part of one of his many movies from New Guinea, before Western civilization infringed on the local culture and rapidly quenched its uniqueness. We continued the evening, just the two of us in my home, with intense conversation, that is, mostly a Gajdusek monologue. His flow of associations was endless and he kept on talking when at midnight I closed the door of the taxi that would take him back to his hotel. His voice has now finally become silent, but, for those of us who have been privileged to learn to know him, we can still hear his intense argumentations ringing in our ears.

In the announcement of his lecture at the Karolinska Institute Gajdusek's address was given as Tromsø, Paris and Amsterdam. The reason for this was that in 1996 he was convicted of pedophilia in the US and after a year in prison and a plea bargain arrangement left his home country for ever. There is no arguing about this. The law applies to everyone and we have learnt during the

20th century that the rights of children should be respected, including, categorically, their right not to be sexually exploited. The fact that earlier cultures and civilizations have judged this kind of relationship differently and that Gajdusek in his engagements to understand the development of children experienced cultures in New Guinea with completely different value systems, provides no excuses. His conviction of a crime cast a dark shadow over the rest of his life and regrettably put him in an occasional defensive mode. For many people it was difficult to keep apart their respect for him as a scientist and their view of him as an individual.

During his time in exile Gajdusek kept in contact with his laboratory in Bethesda. However, in 2004 NIH decided to close this laboratory, which had been in operation for almost 50 years. The last remaining member of the team, Brown, then retired. Gajdusek spent summers mostly in Amsterdam and Paris, but during the winter spell he preferred to live in Tromsø, north of the Arctic circle. In fact he returned to this city some 3–4 times every year. Surprisingly he sought the loneliness this environment offered. Of course he could not resist developing contacts with the aboriginal peoples of this part of Scandinavia, the Sami people or Laplanders. He was fascinated by their religion of Nature and the shamanistic symbols they used.

Gajdusek continued to travel globally outside the US as much as his health and excessive weight allowed. There were frequent visits to China, where he sometimes stayed for longer periods. During these last years he kept in frequent contacts with his many adopted children and their families and also with his many friends, mostly by correspondence. He died alone in a hotel room in Tromsø on December 12, 2008, a time of the year when the sun never rises above the horizon. His ashes were scattered in the Atlantic outside the city at the solstice on June 21, 2009, in the presence of some of his adopted children and friends, from near and far away. The sky was overcast at the event, but when the ashes reached the water the sun broke through the clouds and a large sea eagle passed over the boat. These signs of supernatural events were readily assimilated by the participant people of aboriginal origin. The contrast between the complete absence and omnipresence of sunlight at the two events — his death and the spreading of his ashes — may stand as a symbol of a unique life as full of contrasts as it can ever be.

Gajdusek's close collaborator Brown expressed his feeling of awe and affection for this departed friend in a memoriam[45] in the following way: "Like many men and women of genius, he was bigger than life, with bigger appetites, passions, virtues, and flaws than those who surrounded him … ." In another obituary by Jaap Goudsmit[46] Gajdusek is referred to as "the most outlandish and peripatetic of microbe hunters." Goudsmit retold the story about the stairs with all the names of scientists leading to the attic in Gajdusek's family's home (page 247) and also mentioned that he had left the last step blank for himself. That turns out not to be true. Gajdusek was not that presumptuous. But one should keep in mind, in this situation as in others, that Gajdusek was also a very good story teller.

Science is much more than the joy and enchantment of making major discoveries. In fact most scientists are not granted the privilege and good fortune to make paradigmatic observations. Still they may contribute as a part of a larger enterprise to the accumulation of critical information to be molded into comprehensive knowledge. Like all scientists even the most successful ones meet with frustrations in their work most of the time. There are many blind alleys and challenging of hypotheses that turn out to be flawed. The only way to manage the days of toil, setbacks and disappointments is to stay with the problem, to show perseverance. Gajdusek certainly did this, but I think the other leading character of this chapter, Prusiner, illustrates this even better in his pursuit of science. He was determined to understand the chemical nature of the infectious agent causing scrapie and against all odds he was successful. At the time there was considerable criticism of his ideas, but he was stubborn and held on to them.

Since Prusiner had managed to demonstrate that there is a critical difference between normal healthy prions and those that induce the protein aggregation diseases the most obvious proof of this concept would be to convert the former to the latter in the laboratory. He and his colleagues as well as many other scientists have devoted considerable energy and ingenuity to reaching this goal. Major progress has been made in consolidating the conclusion that the process of conversion is independent of nucleic acid information carrying molecules. Mixing of a small amount of infectious mouse prions with normal mouse brain has led to an increase of infectivity, but no such increase has been achieved when purified reagents have been used. The reason why complete success has not been reached in these endeavors is that the conversion event presumably requires additional components, which as yet have not been identified. Prusiner and other researchers are involved in extensive experiments aimed at explaining the chemical background to strain variations and different disease patterns, as already mentioned[30,31]. In spite of all the interesting results of these efforts it must be concluded that there still remains some pieces to find before the prion jigsaw puzzle is complete.

Prusiner (left) before giving a lecture in 1996 at the Nobel Forum at the Karolinska Institute, in conversation with the author. [Photo by Ulf Sirborn.]

Many of the efforts brought to the prion field are discussed in a highly interesting review that Prusiner has written together with Maclyn McCarty entitled "Discovering DNA encodes heredity and prions are infectious proteins."[47] These two scientists, with McCarty being a critical collaborator of Oswald T. Avery, discussed in the previous chapter, had the opposite challenges. The general problem confronting them was that in science negatives are generally meaningless. Whereas Avery and collaborators had the challenge of proving the absence of proteins in the DNA that they argued to be the transforming principle of pneumococci, Prusiner was trying to prove the opposite, namely the absence of any nucleic acid in his prion protein preparations. Eventually both were proven right, but it took time and the consequences were fundamental. In the case of Prusiner he managed by his perseverance to open a new field of science of much wider application than he could have anticipated. In the future we will see extensive analyses of protein aggregation as an important etiological component in the development of diseases and we will learn amazing things about proteins as information molecules.

References

Chapter 1

1. Espmark, K. (1991) *The Nobel Prize in Literature — A Study of the Criteria behind the Choices.* G. K. Hall & Co., Boston, MA.
2. Holl, K. and Kjelling A. C. (1994) *The Nobel Peace Prize and the Laureates — The Meaning and Acceptance of the Nobel Peace Prize in the Prize Winners' Countries.* Peter Lang GmbH, Europäischer Verlag der Wissenschaften, Frankfurt am Main.
3. Stenersen, O., Libaek, I. and Sveen, A. (2001) *The Nobel Peace Prize. One Hundred Years for Peace.* J. W. Cappelens Forlag AS, Oslo.
4. Schück, H. and Sohlman, R. (1983) *The Legacy of Alfred Nobel.* The Bodley Head, London.
5. Odelberg, W., coordinating editor (1972) *Nobel, the Man and His Prizes.* Elsevier, New York.
6. Fant, K. (1993) *Alfred Bernhard Nobel.* Arcade Publishing, New York.
7. Tolf, R. W. (1976) *The Russian Rockefellers. The Saga of the Nobel Family and the Russian Oil Industry.* Hoover Institution Press, Stanford, CA.
8. Crawford, E. (1984) *The Beginnings of the Nobel Institution. The Science Prizes, 1901–1915.* Cambridge University Press, Cambridge.
9. Ljunggren, B. and Bruyn, G. W. (2002) *The Nobel Prize in Medicine and the Karolinska Institute. The Story of Axel Key and Alfred Nobel.* Karger AG.
10. Jangfeldt, B. (1998) *Svenska vägar till S:t Petersburg (in Swedish).* Wahlström & Widstrand, Stockholm.
11. Crawford, E. (1992) *Nationalism and Internationalism in Science, 1880–1939: Four Studies of the Nobel Population.* Cambridge University Press, Cambridge.
12. Zuckerman, H. (1996) *Scientific Elite. Nobel Laureates in the United States.* New edition with a new introduction. Transaction Publishers, New Brunswick.
13. Hollingsworth, J. R. and Hollingsworth, E. L. (2000) Major discoveries and biomedical research organizations: Perspectives on interdisciplinarity, nurturing leadership and integrated structure and cultures. In *Practising Interdisciplinarity.* Weingart, P. and Stehr, N. (eds.). University of Toronto Press, pp. 215–244.

14. Hollingsworth, J. R. (2004) Institutionalizing excellence in biomedical research: The case of the Rockefeller University. In *Creating a Tradition of Biomedical Research*, Stapleton D. H. (ed.). The Rockefeller University Press, New York, pp. 17–63.
15. Bergström, S. (1977) Opening address at the 1976 Nobel Prize ceremony. In *Les Prix Nobel en 1976*. P. A. Norstedt & Söner, Imprimerie Royale, Stockholm, pp. 17–18.

Chapter 2

1. Cannon, W. B. (1984) *The Way of an Investigator: A Scientist's Experiences in Medical Research*. W. W. Norton, New York.
2. Van Loon, H. W. (1944) *The Arts*. Simon & Schuster, New York.
3. Merton, R. K. (1993) *On the Shoulders of Giants. A Shandean Postscript*. University of Chicago Press, London.
4. Merton, R. K. and Barber, E. (2004) *The Travels and Adventures of Serendipity*. Princeton University Press, Princeton.
5. Roberts, R. M. (1989) *Serendipity: Accidental Discoveries in Science*. John Wiley & Sons.
6. Zuckerman, H. (1996) *Scientific Elite: Nobel Laureates in the United States*. New edition with a new introduction. Transaction Publishers, New Brunswick.
7. Lagerkvist, U. (2005) *The Enigma of Ferment*. World Scientific, Singapore.
8. Blumberg, B. S. (2002) *Hepatitis B: The Hunt for a Killer Virus*. Princeton University Press, Princeton.
9. Kuhn, T. (1962) *The Structure of Scientific Revolutions*. The University of Chicago Press, Chicago.
10. Beadle, G. W. (1966) Biochemical genetics: Some recollections. In *Phage and the Origins of Molecular Biology*. Cairns, J., Stent, G. S. and Watson, J. D. (eds.). Cold Spring Harbor Laboratory Press, New York, pp. 23–32.
11. Feynman, R. P. (1966) The development of the space-time view of quantum electrodynamics. In *Les Prix Nobel en 1965*. P. A. Norstedt & Söner, Imprimerie Royale, Stockholm, pp. 172–191.
12. Hoffmann, R. (1988) Under the surface of the chemical article. *Angewandte Chemie* 27:1593–1602.

Chapter 3

1. van Regenmortel, M. H. V. (2010) Nature of Viruses. In *Desk Encyclopedia of Virology*, Mahy, B. W. J. and Regenmortel, M. H. V. (eds.). Academic Press, San Diego, pp. 19–23.
2. Hughes, S. S. (1977) *The Virus: A History of the Concept*. Heinemann Educational Books, London.
3. Waterson, A. P. and Wilkinson, L. (1978) *An Introduction to the History of Virology*. Cambridge University Press, Cambridge.
4. Fenner, F. and Gibbs, A. (eds.) (1988) *Portraits of Virology: A History of Virology*. Karger, Basel.
5. van Helvoort, T. (1994) History of virus research in the twentieth century: The problem of conceptual continuity. *Hist. Sci.* 32:185–235.

6. Horzinek, M. C. (1997) The birth of virology. *Antonie van Leeuwenhoek* 71:15–20.

7. Calisher, C. H. and Horzinek, M. C. (eds.) (1999) *100 Years of Virology. The Birth and Growth of a Discipline*. Springer Verlag, Vienna.

8. Cairns, J., Stent, G. S. and Watson, J. D. (eds.) (1966) *Phage and the Origins of Molecular Biology*. Cold Spring Harbor Laboratory Press, New York.

9. Olby, R. (1974) *The Path to the Double Helix*. Dover Publications, New York.

10. Lwoff, A. (1957) The concept of virus. *J. Gen. Microbiol.* 17:239–253.

11. Stanley, W. M. (1935) Isolation of a crystalline protein possessing the properties of tobacco-mosaic virus. *Science* 81:644–645.

12. van Helvoort, T. (1991) What is a virus? The case of tobacco mosaic virus. *Stud. Hist. Phil. Sci.* 22:557–588.

13. Creager, A. N. H. (2002). *The Life of a Virus. Tobacco Mosaic Virus as an Experimental Model, 1930–45*. The University of Chicago Press, Chicago.

14. van Regenmortel, M. H. V. (2010). Tobacco mosaic virus. In *Desk Encyclopedia of Virology*. Mahy, B. W. J. and Regenmortel, M. H. V. (eds.). Academic Press, San Diego, pp. 324–340.

15. Klug, A. (1983) From macromolecules to biological assemblies. In *Les Prix Nobel en 1982*. Odelberg, W. (ed.). P. A. Norstedt and Söner, Imprimerie Royale, Stockholm, pp. 89–125.

16. Norrby, E. (2007) Yellow fever and Max Theiler: The only Nobel Prize for a virus vaccine. *J. Exp. Med.* 204:2779–2784.

17. Norrby, E. and Prusiner, B. S. (2007) Polio and Nobel Prizes: Looking back 50 years. *Ann. Neurol.* 61:385–295.

18. Prusiner, S. B. (1998) Prions. In *Les Prix Nobel en 1997*. Frängsmyr, T. (ed.). Almquist & Wiksell International, Stockholm, pp. 268–323.

19. Crawford, E. (1990) The secrecy of Nobel Prize selections in the sciences and its effects on documentation and research. *Proc. Am. Philos. Soc.* 134:408–419.

20. Eriksson-Quensel, I.-B. and Svedberg, T. (1936) Sedimentation and electrophoresis of the tobacco-mosaic virus protein. *J. Am. Chem. Soc.* 58:1863–1867.

21. Reichard, P. (2002) Oswald T. Avery and the Nobel Prize in Medicine. *J. Biol. Chem.* 277:13355–13362.

22. Kay, L. E. (1986) W. M. Stanley's crystallization of the tobacco mosaic virus, 1930–1940. *ISIS* 77:450–472.

23. Bawden, F. C., Pirie, N. W., Bernal, J. D. and Fankuchen, I. (1936) Liquid crystalline substances from virus infected plants. *Nature* 138:1051–1052.

24. Stanley, W. M. (1949) The isolation and properties of crystalline tobacco mosaic virus. In *Les Prix Nobel en 1947* (delayed publication). P. A. Norstedt & Söner, Imprimerie Royale, Stockholm, pp. 196–215.

25. van Helvoort, T. (1992) The controversy between John H. Northrop and Max Delbrück on the formation of bacteriophage: Bacterial synthesis or autonomous multiplication? *Ann. Sci.* 49:545–575.

26. Northrop, J. H. (1948) The preparation of pure enzymes and virus proteins. In *Les Prix Nobel en 1946*. Odelberg, W. (ed.). P. A. Norstedt & Söner, Imprimerie Royale, Stockholm, pp. 193–203.

27. Fraenkel-Conrat, H. (1956) The role of the nucleic acid in the reconstruction of active tobacco mosaic virus. *J. Am Chem. Soc.* 78:882–883.

28. Gierer, A. and Schramm, G. (1956) Infectivity of ribonucleic acid from tobacco mosaic virus. *Nature* 177:702–703.
29. Delbrück, M. (1970) A physicist's renewed look at biology — twenty years later. In *Les Prix Nobel en 1969*. P. A. Norstedt & Söner, Imprimerie Royale, Stockholm, pp. 145–156.
30. Bohr, N. (1933) Light and life. *Nature* 131:421–423, 457–459.
31. van Helvoort, T. (1992) Bacteriological and physiological research styles in the early controversy on the nature of the bacteriophage phenomenon. *Med. Hist.* 36:243–270.
32. Hershey, A. D. and Chase, M. (1952) Independent functions of viral protein and nucleic acid in growth of bacteriophage. *J. Gen. Physiol.* 36:39–56.
33. Lwoff, A. (1966) Interactions entre virus, cellule et organisme. In *Les Prix Nobel en 1965*. P. A. Norstedt & Söner, Imprimerie Royale, Stockholm, pp. 233–243.
34. Norrby, E., Magnusson, P., Falksveden, L. G. and Grönberg, M. (1964) Separation of measles virus in CsCl gradients. II. Studies on the large and small hemagglutinin. *Arch. Ges. Virusforsch.* 14:462–473.
35. Lederberg, J. (1959) A view of genetics. In *Les Prix Nobel en 1958*. P. A. Norstedt & Söner, Imprimerie Royale, Stockholm, pp. 170–189.
36. Rivers, T. M. (ed.) (1932) The nature of viruses. *Physiol. Rev.* 12:423–452.
37. Hoyle, L. (1968) The influenza viruses. *Virology Monographs* 4:15–54.
38. Gard, S. (1955) Presentation speech to the Nobel Prize in Physiology or Medicine 1954. In *Les Prix Nobel en 1954*. P. A. Norstedt & Söner, Imprimerie Royale, Stockholm, pp. 33–37.
39. Burnet, F. M. and Andrewes, C. H. (1933) Über die Natur der filtrierbaren Vira. *Zentralbl. Bakt. Parasit. Infektionsk. Abt I, Orig.* 130:161–183.
40. Dubos, R. J. (1976) *The Professor, the Institute and DNA. Oswald T. Avery: His Life and Scientific Achievements*. The Rockefeller University Press, New York, pp. 156–157.
41. Burnet, F. M. (1955) *Viruses and Man, 2nd edition*. Penguin Books, Harmondworth, Middlesex.
42. Westgren, A. (1972) The Prize in Chemistry. In *Nobel, the Man and His Prizes. 3rd ed.* Odelberg, W. (ed.). Elsevier, New York, pp. 281–385.
43. Hershey, A. D. (1970) Idiosyncrasies of DNA structure. In *Les Prix Nobel en 1969*. Odelberg, W. (ed.). P. A. Norstedt & Söner, Imprimerie Royale, Stockholm, pp. 157–163.
44. Raoult, D., Audic, S., Robert, C. *et al.* (2004) The 1.2-megabase genome sequence of mimivirus. *Science* 306: 1344–1350.
45. Villarreal, L. P. (2005) *Viruses and the Evolution of Life*. ASM Press, Washington.

Chapter 4

1. Crawford, E. (1990) The secrecy of Nobel Prize selections in the sciences and its effects on documentation and research. *Proc. Am. Phil. Soc.* 134:408–419.
2. Powell, J. H. (1993) *Bring Out Your Dead. The Great Plague of Philadelphia in 1793*. University of Pennsylvania Press, 1949 (reprinted with a new Foreword).
3. Bowers, J. Z. and King, E. E. (1981) The conquest of yellow fever. *J. Med. Soc.* 78:539–541.

4. Strode, G. K. (ed.) (1951) *Yellow Fever*. McGraw-Hill, New York.
5. Monath, T. P. (2004) Yellow fever: An update. *Lancet Infect. Dis.* 1:11–20.
6. Lefeuvre, A., Marianneau, P. and Deubel, V. (2004) Current assessment of yellow fever and yellow fever vaccine. *Curr. Infect. Dis. Rep.* 6:96–104.
7. Monath, T. P. (1991) Yellow fever: Victor, Victoria? Conqueror, conquest? Epidemics and research in the last forty years and prospects for the future. *Amer. J. Trop. Med.* 45:1–43.
8. Reed, W. and Carroll, J. (1902) The etiology of yellow fever. *Amer. Med.* 3:301–305.
9. Plesset, I. (1980) *Noguchi and His Patrons*. Fairleigh Dickinson University Press, Madison, New Jersey.
10. Stokes, A., Bauer, J. H. and Hudson, N. P. (1928) Transmission of yellow fever to *Macacas rhesus*, preliminary note. *J. Amer. Med. Ass.* 90:253–254.
11. Bauer, J. H. and Mahaffy, A. F. (1930) Studies of the filtrability of yellow fever virus. *Am. J. Hyg.* 12:175–195.
12. Theiler, M. and Gard, S. (1940) Encephalomyelitis of mice. I. Characterization and pathogenesis of the virus. *J. Exp. Med.* 72:49–67.
13. Theiler, M. and Gard, S. (1940) Encephalomyelitis of mice. III. Epidemiology. *J. Exp. Med.* 72: 79–90.
14. Fox, J. P. and Gard, S. (1940) Preservation of yellow fever virus. *Am. J. Trop. Med.* 20:447–451.
15. Hollingsworth, J. R. (2004) Excellence in biomedical research: The case of Rockefeller University. In *Creating a Tradition of Biomedical Research: The Rockefeller University Centennial History Conference*. Darwin H. Stapleton (ed.). Rockefeller University Press, New York, pp. 17–63.
16. Zuckerman, H. (1996, new edition). *Scientific Elite*. Transaction Publishers, New Brunswick.
17. Zuckerman, H. (1963) Interview with M. Theiler on October 21, 1963. Oral History Research Office, Columbia University, New York.
18. Mathis, C., Sellards, A. W. and Laigret, J. (1928) Sensibilité du *Macacus rhesus* au virus fievre jaune. *Compt. Rend. Acad. Sci.* 186:604–606.
19. Theiler, M. and Sellards, A. W. (1926) Relationship of *L. icterohaemorrhagiae* and *L. icteroides* as determined by Pfeiffer phenomenon in guinea pigs. *Am. J. Trop. Med.* 6:383–402.
20. Theiler, M. and Sellards, A. W. (1928) Immunological relationship of yellow fever as it occurs in West Africa and South America. *Ann. Trop. Med.* 22:449–460.
21. Theiler, M. (1930) Susceptibility of white mice to the virus of yellow fever. *Science* 71:367.
22. Theiler, M. (1930) Studies on the action of yellow fever virus in mice. *Ann. Trop. Med. Parasit.* 24:249–272.
23. Lemmel, B. (2001) Nomination and selection of the Nobel laureates. In *The Nobel Prize: The First 100 Years*. Wallin-Levinovitz, A. and Ringertz, N. (eds.). Imperial College Press, London, pp. 25–28.
24. Sabin, A. B. (1949) Antigenic relationships of dengue and yellow fever viruses with those of West-Nile and Japanese B encephalitis. *Fed. Proc.* 8:410.
25. Theiler, M. (1933) A yellow fever protection test in mice by intracerebral injection. *Amer. Trop. Med. Parasit.* 25:57–77.

26. Theiler, M. and Haagen, E. (1932) Studies of yellow fever virus in tissue culture. *Proc. Soc. Exp. Biol. Med.* 29:435–436.

27. Theiler, M. and Haagen, E. (1932) Untersuchungen über das Verhalten des Gelbfiebervirus in der Gewebekultur. *Zbl. Bakt. I Orig.* 125:145–158.

28. Theiler, M. and Whitman, L. (1935) The danger with vaccination with neurotropic yellow fever virus alone. *Bull. mens. de l'Offic. Inst. Hyg. Publ.* 27:1342–1347.

29. Theiler, M. and Whitman, L. (1935) Quantitative studies of the virus and immune serum used in vaccination against yellow fever. *Amer. J. Trop. Med.* 15:347–356.

30. Lloyd, W., Theiler, M. and Ricci, N. I. (1936) Modification of yellow fever virus by cultivation in tissues *in vitro. Trans. Roy. Soc. Med. Hyg.* 29:481–529.

31. Theiler, M. and Smith, H. H. (1937) The effect of prolonged cultivation *in vitro* upon the pathogenicity of yellow fever virus. *J. Exp. Med.* 65:767–786.

32. Theiler, M. and Smith, H. H. (1937) The use of yellow fever modified by *in vitro* cultivation for human immunization. *J. Exp. Med.* 65:787–800.

33. WHO (2003) Yellow fever vaccine: WHO position paper. *Wkly. Epidemiol. Rec.* 78:349–360.

34. Fox, J. P., Manso, C., Penna, H. A. and Para, M. (1942) Observations on the occurrence of icterus in Brazil following vaccination against yellow fever. *Am. J. Hyg.* 36:68–114.

35. Sawyer, W. A., Meyer, K. F., Eaton, M. D. *et al.* (1944) Jaundice in army personnel in the western region of the United States and its relation to vaccination against yellow fever (Parts II, III and IV). *Am. J. Hyg.* 40:35–104.

36. Bartholomew, J. R. (2002) Katsusaburo Yamagiwa's Nobel candidacy: Physiology or medicine in the 1920s. In *Historical Studies in the Nobel Archives. The Prizes in Science and Medicine.* Crawford, E. (ed.). Universal Academy Press, Tokyo, pp. 107–131.

37. Norrby, E. and Prusiner, S. B. (2007). Polio and Nobel Prizes: Looking back fifty years. *Ann. Neurol.* 61:385–395.

38. Offit, P. A. (2007) *Vaccinated. One Man's Quest to Defeat the World's Deadliest Diseases.* Harper Collins, New York.

39. Allen, A. (2007) Vaccine. *The Controversial Story of Medicine's Greatest Lifesaver.* W. W. Norton, New York, pp. 221–222 ff.

40. Kurth, R. (2005). Obituary. Maurice R. Hilleman (1919–2005). *Nature* 434:1083.

41. Norrby, E. (2007) Yellow fever and Max Theiler: The only Nobel Prize for a virus vaccine. *J. Exp. Med.* 204:2779–2784.

42. Bergstrand, H. (1952) Introductory speech to the Nobel Prize for Physiology or Medicine 1951. In *Les Prix Nobel en 1951.* P. A. Norstedt & Söner, Imprimerie Royale. Stockholm, pp. 40–43.

43. Rice, C. M., Lenches, E. M., Eddy, S. R. *et al.* (1985) Nucleotide sequence of yellow fever virus: Implications for flavivirus gene expression and evolution. *Science* 229:726–735.

44. Van Epps, H. L. (2005) Broadening the horizons for yellow fever: New uses for an old vaccine. *J. Exp. Med.* 201:165–168.

45. Chambers, T. J., Nestorowicz, A., Mason, P. W. and Rice, C. M. (1999) Yellow fever/Japanese encephalitis chimeric viruses: Construction and biological properties. *J. Virol.* 73:3095–3101.

46. Monath, T. P., Guirakhoo, F., Nichols, R. *et al.* (2003) Chimeric live, attenuated vaccine against Japanese encephalitis (ChimeriVax-JE): Phase 2 clinical trials for safety and immunogenicity, effect of virus dose and schedule, and memory response to challenge with Japanese encephalitis antigen. *J. Infect. Dis.* 188:1213–1230.

47. Guirakhoo, F., Pugachev, K., Zhang, Z. *et al.* (2004) Safety and efficacy of chimeric yellow fever-dengue virus tetravalent vaccine formulations in non-human primates. *J. Virol.* 78:4761–4775.

48. Bonaldo, M. C., Garratt, M. S., Caufour, P. S. *et al.* (2002) Surface expression of an immunodominant malaria protein B cell epitope by yellow fever virus. *J. Mol. Biol.* 315:873–885.

49. Tao, D., Barba-Spaeth, G., Rai, U. *et al.* (2005) Yellow fever 17D as a vaccine vector for microbial CTL epitopes: Protection in a rodent malaria model. *J. Exp. Med.* 201:201–209.

50. Theiler, M. (1952) Speech at the Nobel Banquet. In *Les Prix Nobel en 1951*. P. A. Norstedt & Söner, Imprimerie Royale, Stockholm, pp. 63–64.

Chapter 5

1. Salk, J. E., Bazeley, P. L., Bennett, B. L. *et al.* (1954) Studies in human subjects on active immunization against poliomyelitis. II. A practical means for inducing and maintaining antibody formation. *Am. J. Public Health.* 44:994–1009.

2. Salk, J. E., Youngner, J. S. and Ward, E. N. (1954) Use of color change of phenol red as the indicator in titrating poliomyelitis virus or its antibody in a tissue-culture system. *Am. J. Hyg.* 60:214–230.

3. Salk, J. E. (1955) Considerations in the preparation and use of poliomyelitis virus vaccine. *J. Am. Med. Assoc.* 158:1239–1248.

4. Koprowski, H., Jervis, G. A. and Norton, T. W. (1952) Immune responses in human volunteers upon oral administration of a rodent-adapted strain of poliomyelitis virus. *Am. J. Hyg.* 55:108–124.

5. Cox, H. R., Jervis, G. A., Koprowski, H. *et al.* (1956) Immunization of humans with a chick embryo adapted strain of MEF1 poliomyelitis virus. *J. Immunol.* 77:123–131.

6. Koprowski, H. (2006) First decade (1950–1960) of studies and trials with the polio vaccine. *Biologicals* 34:81–86.

7. Sabin, A. B. (1957) Present status of attenuated live virus poliomyelitis vaccine. *Bull. N.Y. Acad. Med.* 33:17–39.

8. Francis, T, Jr., Napier, J. A., Voight, R. B. *et al.* (1957) Evaluation of the 1954 Field Trial of Poliomyelitis Vaccine: Final Report. University of Michigan, Ann Arbor.

9. Norrby, E. and Prusiner, S. B. (2007) Polio and Nobel Prizes: Looking back 50 years. *Ann. Neurol.* 61:385–395.

10. Paul, J. R. (1971) *A History of Poliomyelitis*. Yale University Press, New Haven.

11. Nathanson, N. and Martin, J. R. (1979) The epidemiology of poliomyelitis: Enigmas surrounding its appearance, epidemicity, and disappearance. *Am. J. Epidemiol.* 110:672–692.

12. Landsteiner, K. and Popper, E. (1908) Mikroskopische preparate von einen menschlichen und zwei affenmunckenmarken. *Wien. Klin. Wochenschr.* 21:1830.

13. Brodie, M., Goldberg, S. A. and Stanley, P. (1935) Transmission of the virus of poliomyelitis to mice. *Science* 29:319–320.
14. Armstrong, C. (1939) The experimental transmission of poliomyelitis to the Eastern cotton rat, *Sigmodon hispidus hispidus*. *Public Health Rep.* 54:1719–1721.
15. Armstrong, C. (1939) Successful transfer of the Lansing strain of poliomyelitis virus from the cotton rat to the white mouse. *Public Health Rep.* 54:2302–2305.
16. Enders, J. F., Weller, T. H. and Robbins, F. C. (1949) Cultivation of the Lansing strain of poliomyelitis virus in cultures of various human embryonic tissues. *Science* 109:85–87.
17. Weller, T. H., Robbins, F. C. and Enders, J. F. (1949) Cultivation of poliomyelitis virus in cultures of human foreskin and embryonic tissues. *Proc. Soc. Exp. Biol. Med.* 72:153–155.
18. Robbins, F. C., Enders, J. F. and Weller, T. H. (1950) Cytopathogenic effect of poliomyelitis viruses *in vitro* on human embryonic tissues. *Proc. Soc. Exp. Biol. Med.* 75:370–374.
19. Robbins, F. C., Weller, T. H. and Enders, J. F. (1952) Studies on the cultivation of poliomyelitis viruses in tissue culture. II. The propagation of the poliomyelitis viruses in roller-tube cultures of various human tissues. *J. Immunol.* 69:673–694.
20. Koprowski, H., Norton, T. W., Jervis, G. A. *et al.* (1956) Clinical investigations on attenuated strains of poliomyelitis virus; use as a method of immunization of children with living virus. *J. Am. Med. Assoc.* 160:954–966.
21. Koprowski, H. (1957) Vaccination with modified active viruses. In Congress, I. P. (ed.), *Poliomyelitis: Papers and discussions presented at the Fourth International Poliomyelitis Conference.* J. B. Lippincott, Philadelphia, pp. 112–123.
22. Flexner, S. and Lewis, P. A. (1910) Experimental epidemic poliomyelitis in monkeys. *J. Exp. Med.* 12:227–255.
23. Flexner, S. and Clark, P. F. (1912) A note on the mode of infection in epidemic poliomyelitis. *Proc. Soc. Exp. Biol. Med.* 10:1–4.
24. Sabin, A. B. and Olitsky, P. K. (1936) Cultivation of poliomyelitis virus *in vitro* in human embryonic nervous tissue. *Proc. Soc. Exp. Biol. Med.* 34:357–359.
25. Feller, A. E., Enders, J. F. and Weller, T. H. (1940) The prolonged coexistence of vaccinia virus in high titre and living cells in roller-tube cultures of chick embryonic tissues. *J. Exp. Med.* 72:367–388.
26. Weller, T. H. and Enders, J. F. (1949) Propagation of hemagglutinin by mumps and influenza A viruses in suspended cell tissue cultures. *Proc. Soc. Exp. Biol. Med.* 69:124–128.
27. Weller, T. H. (1953) Serial propagation *in vitro* of agents producing inclusion bodies derived from varicella and herpes zoster. *Proc. Soc. Exp. Biol. Med.* 83:340–346.
28. Enders, J. F. and Peebles, T. C. (1954) Propagation in tissue cultures of cytopathogenic agents from patients with measles. *Proc. Soc. Exp. Biol. Med.* 86:277–286.
29. Weller, T. H. (2004) *Growing Pathogens in Tissue Cultures: Fifty Years in Academic Tropical Medicine, Pediatrics, and Virology.* Boston Medical Library, Science History Publications, USA.
30. Zuckerman, H. (1964) Interview with Frederick C. Robbins on April 2, 1964. Oral History Research Office, Columbia University, New York.

31. Hargittai, I. (2002) *Candid Science II: Conversations with Famous Biomedical Scientists.* Imperial College Press, London.
32. Theiler, M. and Gard, S. (1940) Encephalomyelitis of mice. I. Characteristics and pathogenesis of the virus. *J. Exp. Med.* 72:49–67.
33. Theiler, M. and Gard, S. (1940) Encephalomyelitis of mice. III. Epidemiology. *J. Exp. Med.* 72:79–90.
34. Gard, S. (1943) Purification of poliomyelitis virus. Thesis, Institute of Physical Chemistry, Department of Hygiene and Bacteriology, University of Uppsala.
35. Salk, J. E., Krech, U., Youngner, J. S. *et al.* (1954) Formaldehyde treatment and safety testing of experimental poliomyelitis vaccines. *Am. J. Public Health.* 44:563–570.
36. Gard, S. (1956) Aspects on production and control of formol-treated poliovirus vaccines. *European Association of Poliomyelitis, IV Symposium*, pp. 22–25.
37. Gard, S., Lycke, E., Olin, G. and Wesslen, T. (1957) Inactivation of poliomyelitis virus by formaldehyde. *Arch. Gesamte Virusforsch.* 7:125–135.
38. Gard, S. and Lycke, E. (1957) Inactivation of poliovirus by formaldehyde; analysis of inactivation curves. *Arch. Gesamte Virusforsch.* 7:471–482.
39. Lycke, E. (1958) Studies of the inactivation of poliomyelitis virus by formaldehyde; inactivation of partially purified virus material and the effect upon the rate of inactivation by addition of glycine. *Arch. Gesamte Virusforsch.* 8:23–41.
40. Salk, J. E. and Gori, J. B. (1960) A review of theoretical, experimental, and practical considerations in the use of formaldehyde for the inactivation of poliovirus. *Ann. NY Acad. Sci.* 83:609–637.
41. Nathanson, N. and Langmuir, A. D. (1963) The Cutter Incident. Poliomyelitis following formaldehyde-inactivated poliovirus vaccination in the United States during the Spring of 1955. I. Background. *Am. J. Hyg.* 78:16–28.
42. Nathanson, N. and Langmuir, A. D. (1963) The Cutter Incident. Poliomyelitis following formaldehyde-inactivated poliovirus vaccination in the United States during the Spring of 1955. II. Relationship of poliomyelitis to Cutter vaccine. *Am. J. Hyg.* 78:29–60.
43. Nathanson, N. and Langmuir, A. D. (1963) The Cutter Incident. Poliomyelitis following formaldehyde-inactivated poliovirus vaccination in the United States during the Spring of 1955. III. Comparison of the clinical character of vaccinated and contact cases occurring after use of high rate lots of Cutter vaccine. *Am. J. Hyg.* 78:61–81.
44. Offit, P. A. (2005) *The Cutter Incident: How America's First Polio Vaccine Led to the Growing Vaccine Crisis.* Yale University Press, New Haven.
45. Gard, S. (1957) Chemical inactivation of viruses. *Ciba Foundation Symposium on the Nature of Viruses.* Churchill, London.
46. Böttiger, M. (1966) Studies on immunization with inactivated and live poliovirus vaccines. *Acta Paediatr. Scand. Suppl.* 164:1–42.
47. Böttiger, M., Arro, L., Lundbäck, H. and Salenstedt, C. R. (1966) The immune response to vaccination with inactivated poliovirus vaccine in Sweden. *Acta Pathol. Microbiol. Scand.* 66:239–256.
48. Böttiger, M., Lycke, E., Melén, B. and Wrange, G. (1958) Inactivation of poliomyelitis virus by formaldehyde; incubation time in tissue culture of formalin- treated virus. *Arch. Gesamte Virusforsch.* 8:259–266.

49. Böttiger, M. (1981) Experiences of vaccination with inactivated poliovirus vaccine in Sweden. *Dev. Biol. Stand.* 47:227–232.

50. Enders, J. F., Robbins, F. C. and Weller, T. H. (1955) The cultivation of the poliomyelitis viruses in tissue culture. In *Les Prix Nobel en 1954*. P. A. Norstedt & Söner, Stockholm, pp. 100–118.

51. Nathanson, N. and Bodian, D. (1961) Experimental poliomyelitis following intramuscular virus injection. II. Viremia and the effect of antibody. *Bull. Johns Hopkins Hosp.* 108:320–333.

52. Nathanson, N. and Bodian, D. (1961) Experimental poliomyelitis following intramuscular virus injection. I. The effect of neural block on a neurotropic and a pantropic strain. *Bull. Johns Hopkins Hosp.* 108:308–319.

53. Rosen, F. S. (2004) Isolation of poliovirus — John Enders and the Nobel Prize. *N. Engl. J. Med.* 351:1481–1483.

54. Roca-Garcia, M., Moyer, A. W. and Cox, H. R. (1952) Poliomyelitis. II. Propagation of MEF1 strain of poliomyelitis virus in developing chick embryo by yolk sac inoculation. *Proc. Soc. Exp. Biol. Med.* 81:519–525.

55. Cabasso, V. J., Stebbins, M. R., Dutcher, R. M. *et al.* (1952) Poliomyelitis. III. Propagation of MEF1 strain of poliomyelitis virus in developing chick embryo by allantoic cavity inoculation. *Proc. Soc. Exp. Biol. Med.* 81:525–529.

56. Lemmel, B. (2001) Nomination and selection of the Nobel Laureates. In Wallin-Levinovitz, A. and Ringertz, N. (eds.), *The Nobel Prize: The First 100 Years*. London: Imperial College Press, pp. 25–28.

57. Crawford, E. (1990) The secrecy of Nobel Prize selections in the sciences and its effect on documentation and research. *Proc. Am. Philos. Soc.* 134:408–419.

58. Special to *The New York Times* (1954). 3 U.S. doctors win Nobel award for work in growing polio virus. *The New York Times*, New York.

59. Enders, J. F., Weller, T. H. and Robbins, F. C. (1952) Alterations in pathogenicity for monkeys of Brunhilde strain of poliomyelitis virus following cultivation in human tissues. *Fed. Proc.* 11:467.

60. Enders, J. F., Katz, S. L., Milovanovic, M. V. and Holloway, A. (1960) Studies on an attenuated measles-virus vaccine. I. Development and preparations of the vaccine: Technics for assay of effects of vaccination. *N. Engl. J. Med.* 263:153–159.

61. Salk, J. E. (1953) Studies in human subjects on active immunization against poliomyelitis. I. A preliminary report of experiments in progress. *J. Am. Med. Assoc.* 151:1081–1098.

62. Dulbecco, R. and Vogt, M. (1954) Plaque formation and isolation of pure lines with poliomyelitis viruses. *J. Exp. Med.* 99:167–182.

63. Rous, P. and Jones, F. S. (1916) A method for obtaining suspensions of living cells from the fixed tissues, and for the plating out of individual cells. *J. Exp. Med.* 23:549–555.

64. Youngner, J. S. (1954) Monolayer tissue cultures. II. Poliomyelitis virus assay in roller-tube cultures of trypsin-dispersed monkey kidney. *Proc. Soc. Exp. Biol. Med.* 85:527–530.

65. Youngner, J. S. (1954) Monolayer tissue cultures. I. Preparation and standardization of suspensions of trypsin-dispersed monkey kidney cells. *Proc. Soc. Exp. Biol. Med.* 85:202–205.

66. Sabin, A. B., Hennessen, W. A. and Winsser, J. (1954) Studies on variants of poliomyelitis virus. I. Experimental segregation and properties of avirulent variants of three immunologic types. *J. Exp. Med.* 99:551–576.

67. Gard, S. (1955) Introductory speech to the Nobel Prize for Physiology or Medicine 1954. In *Les Prix Nobel en 1954*. P. A. Norstedt & Söner, Stockholm, pp. 38–42.

Chapter 6

1. Liljestrand, G. (1960) Karolinska Institutet och Nobelprisen (in Swedish). In *Karolinska Mediko-Kirurgiska Institutets Historia 1910–1960*, Kapitel XI, Almqvist & Wiksell, Stockholm, pp. 536–590.

2. Liljestrand, G. (1972) The Prize in Physiology or Medicine. In *Nobel, The Man and His Prizes, 3rd edition*. Odelberg, W. (ed). Elsevier, New York, pp. 139–278.

3. Norrby, E. (2007) Yellow fever and Max Theiler: The only Nobel Prize for a virus vaccine. *J. Exp. Med.* 204:2779–2784.

4. Norrby, E. and Prusiner, S. B. (2007) Polio and Nobel Prizes: Looking back 50 years. *Ann. Neurol.* 61:385–395.

5. Lagerkvist, U. (2003) *Pioneers of Microbiology and the Nobel Prize*. World Scientific, Singapore.

6. Salamon-Bayet, C. (1982) Bacteriology and Nobel Prize selections. 1901–1920. In *Science, Technology and Society in the Time of Alfred Nobel*. Bernard, C. G., Crawford, E. and Sörbom, P. (eds.). Pergamon Press, Oxford, pp. 377–400.

7. Hessenbruch, A. and Petersen, F. (2001) Niels Finsen (Physiology or Medicine 1903) "Banishing darkness and disease". In *The History of Thirteen Danish Nobel Prizes. Neighbouring Nobel*. Nielsen, H. and Nielsen, K. (eds.). Aarhus University Press, pp. 393–429.

8. Crawford, E. (1996) *Arrhenius. From Ionic Theory to the Greenhouse Effect*. Watson Publishing International, Canton, MA.

9. Åkerman, J. (1913) Introductory speech to the Nobel Prize in Physiology or Medicine 1912. In *Les Prix Nobel en 1912*. P. A. Norstedt & Söner, Imprimerie Royal, Stockholm, pp. 25–29.

10. Carrel, A. (1913) Suture of blood-vessels and transplantation of organs. In *Les Prix Nobel en 1912*. P. A. Norstedt & Söner, Imprimerie Royal, Stockholm, pp. 15–25.

11. McKellar, S. (2004) Innovation in modern surgery: Alexis Carrel and blood vessel repair. In *Creating a Tradition of Biomedical Research*. Stapleton, D. H. (ed.). Rockefeller University Press, New York, pp. 135–150.

12. Edwards, W. S. and Edwards, P. D. (1974) *Alexis Carrel: Visionary surgeon*. Charles C. Thomas, Springfield, IL.

13. Reggiani, A. H. (2006) *God's Eugenicist: Alexis Carrel and the Sociobiology of Decline*. Berghahn Books, New York.

14. Friedman, D. M. (2008) *The Immortalists: Charles Lindbergh, Dr Alexis Carrel, and Their Daring Quest to Live Forever*. Tantor Media, Connecticut.

15. Newton, J. D. (1989) *Uncommon Friends: Life with Thomas Edison, Henry Ford, Harvey Firestone, Alexis Carrel, and Charles Lindbergh*. Harcourt, Florida.

16. Landecker, H. (2004) Building "A new type of body in which to grow a cell":

Tissue culture at the Rockefeller Institute, 1910–1914. In *Creating a Tradition of Biomedical Research*. Stapleton, D. H. (ed.). Rockefeller University Press, New York, pp. 151–174.

17. Takahashi, A. (2004) Hideyo Noguchi, the pursuit of immunity and the persistence of fame: A reappraisal. In *Creating a Tradition of Biomedical Research*. Stapleton, D. H. (ed.). Rockefeller University Press, New York, pp. 227–239.

18. Carrel, A. (1948) *Man the Unknown*. Penguin Books, West Drayton.

19. Nielsen, A. K. (2001) August Krogh (Physiology or Medicine 1920) "Scientist explains blushing of girls." In *The History of Thirteen Danish Nobel Prizes. Neighbouring Nobel*. Nielsen, H. and Nielsen K. (eds.). Aarhus University Press, Aarhus, pp. 430–460.

20. Nielsen, H. and Nielsen, K. (2002) Neighbouring Nobel: A look at the Danish laureates. In *Historical Studies in the Nobel Archives. The Prizes in Science and Medicine*. Crawford, E. (ed.). Universal Academy Press, Tokyo, pp. 133–154.

21. Nielsen, A. K. and Thorling, E. B. (2002) Johannes Fibiger (Physiology or Medicine 1926). Backing the wrong horse? In *The History of Thirteen Danish Nobel Prizes. Neighbouring Nobel*. Nielsen, H. and Nielsen, K. (eds.). Aarhus University Press, Aarhus, pp. 461–493.

22. Stolt, C. M., Klein, G. and Jansson, A. T. R. (2004) An analysis of a wrong Nobel Prize — Johannes Fibiger, 1926: A study in the Nobel archives. *Adv. Cancer Res.* 92:1–12.

23. Needham, D. M. (1971) *Machina Carnis: The Biochemistry of Muscular Contraction in Its Historical Development*. Cambridge University Press, Cambridge.

24. Johansson, J. E. (1924) Introductory speech to the Nobel Prize in Physiology or Medicine 1922. In *Les Prix Nobel en 1923*. P. A. Norstedt & Söner, Imprimerie Royal, Stockholm, pp. 31–37.

25. Bliss, M. (1982) *The Discovery of Insulin*. McClelland and Stewart Limited, Toronto.

26. Sjöquist, J. (1924) Introductory speech to the Nobel Prize in Physiology or Medicine 1923. In *Les Prix Nobel en 1923*. P. A. Norstedt & Söner, Imprimerie Royal, Stockholm, pp. 46–50.

27. MacLeod, J. J. R. (1926) The physiology of insulin and its source in the animal body. In *Les Prix Nobel en 1924–1925. Les conferénces Nobel*. P. A. Norstedt & Söner, Imprimerie Royal, Stockholm, pp. 1–12.

28. Banting, F. G. (1926) Diabetes and insulin. In *Les Prix Nobel en 1924–1925. Les conferénces Nobel*. P. A. Norstedt & Söner, Imprimerie Royal, Stockholm, pp. 1–20.

29. Bliss, M. (1993) Rewriting medical history: Charles Best and the Banting and Best myth. *J. Med. Hist. All. Sci.* 48:253–274.

30. Kragh, H. and Möller, M. K. (2001) Henrik Dam (Physiology or Medicine 1943). The anonymous laureate. *The History of Thirteen Danish Nobel Prizes. Neighbouring Nobel*. Nielsen, H. and Nielsen, K. (eds.). Aarhus University Press, Aarhus, pp. 494–522.

31. Liljestrand, G. (1947) Introductory speech to the Nobel Prize in Physiology or Medicine 1945. In *Les Prix Nobel en 1945*. P. A. Norstedt & Söner, Imprimerie Royal, Stockholm, pp. 31–36.

32. Macfarlane, G. (1979) *Howard Florey: Making of a Great Scientist*. Oxford University Press, Oxford.

33. Stapleton, D. H. (2005) A lost chapter in the early history of DDT. The development

of anti-typhus technologies by the Rockefeller Foundation's Louse laboratory, 1942–1944. *Technology and Culture* 46:513–540.

34. Perkins, J. H. (1978) Reshaping technology in wartime: The effect of military goals on entomological research and insect control practises. *Technology and Culture* 19:169–182.

35. Fischer, G. (1949) Introductory speech to the Nobel Prize in Physiology or Medicine 1948. In *Les Prix Nobel en 1948*. P. A. Norstedt & Söner, Imprimerie Royal, Stockholm, pp. 35–38.

36. Müller, P. (1949) Dichlordiphenyltrichloräthan und neuere insektizide. In *Les Prix Nobel en 1948*. P. A. Norstedt & Söner, Imprimerie Royal, Stockholm, pp. 122–132.

37. Carson, R. (1962) *Silent Spring*. Houghton Mifflin Harcourt, New York.

38. Stapleton, D. H. (2000) The short-lived miracle of DDT. *American Heritage of Invention and Technology* 15:34–41.

39. Lundberg, I. E., Grundtman, C., Larsson, E. and Klareskog, L. (2004) Corticosteroid — from an idea to clinical use. *Best Practise and Research Clinical Rheumatology* 18:7–19.

40. Liljestrand, G. (1951) Introductory speech to the Nobel Prize in Physiology or Medicine 1950. In *Les Prix Nobel en 1950*. P. A. Norstedt & Söner, Imprimerie Royal, Stockholm, pp. 34–40.

41. Ragnarsson, U. (2007) The Nobel trail of Vincent du Vigneaud. *J. Peptide Sci.* 13:431–433.

42. Norrby, E. (2008) Nobel Prizes and the emerging virus concept. *Arch. Virol.* 153:1109–1123.

43. Lederberg, J. (1959) A view of genetics. In *Les Prix Nobel en 1958*. P. A. Norstedt & Söner, Imprimerie Royale, Stockholm, pp. 170–189.

Chapter 7

1. Monod, J. (1971) *Chance and Necessity: An Essay of the Natural Philosophy of Modern Biology*, Penguin Books, St. Ives.

2. Olby, R. (1974) *The Path to the Double Helix. The Discovery of DNA*. Dover Publications, New York.

3. Portugal, F. H. and Cohen, J. S. (1977) *A Century of DNA*. The MIT Press, Cambridge, MA.

4. Jacob, F. (1976) *The Logic of Life. A History of Heredity*. Vintage Books, Random House, New York.

5. Brock, T. D. (1990) *The Emergence of Bacterial Genetics*. Cold Spring Harbor Laboratory Press, New York.

6. Kay, L. E. (1993) *The Molecular Vision of Life*. Oxford University Press, New York.

7. Pollack, R. (1994) *Signs of Life. The Language and Meaning of DNA*. Houghton Mifflin, New York.

8. Judson, H. F. (1996) *The Eight Days of Creation: Makers of the Revolution in Biology*. Cold Spring Harbor Laboratory Press.

9. Reichard, P. (2002) Osvald T. Avery and the Nobel Prize in Medicine. *J. Biol. Chem.* 277:13355–13362.

10. Watson, J. D. (1980) *The Double Helix. A Personal Account of the Discovery of the*

Structure of DNA. Atheneum, New York.

11. Crick, F. (1990) *What Mad Pursuit: A Personal View of Scientific Discovery*. Basic Books, New York.

12. Wilkins, M. (2003) *The Third Man of the Double Helix*. Oxford University Press.

13. Dahm, R. (2008). Discovering DNA: Friedrich Miescher and the early years of nucleic acid research. *Hum. Genet.* 122:565–581.

14. Lagerkvist, U. (1998) *DNA Pioneers and Their Legacy*. Yale University Press, New Haven

15. Mörner, K. A. H. (1911) Introductory speech to the 1910 Nobel Prize in Physiology or Medicine. In *Les Prix Nobel en 1910*. P. A. Norstedt & Söner, Imprimerie Royal, Stockholm, pp. 22–27.

16. Wilson, E. B. (1896) *The Cell in Development and Inheritance. 1st edition.* Macmillan, New York.

17. Avery, O. T., MacLeod, C. M., and McCarty, M. (1944) Studies on the chemical nature of the substance inducing transformation of pneumococcal types. Induction of transformation by a deoxyribonucleic acid fraction isolated from pneumococcus type. III. *J. Exp. Med.* 79:137–157.

18. Wyatt, H. V. (1972) When does information become knowledge? *Nature* 235:86–89.

19. Dubos, R. J. (1976) The *Professor, the Institute and DNA*. Rockefeller University Press, New York.

20. Deichmann, U. (2004) Early responses to Avery et al.'s paper on DNA as hereditary material. *Hist. Stud. Phys. Biol. Sci.* 34:207–232.

21. Prusiner, S. B. and McCarty, M. (2006) Discovering DNA encodes heredity and prions are infectious proteins. *Ann. Rev. Genet.* 40:25–45.

22. Watson, A. D. and Crick, F. H. C. (1953) Molecular structure of nucleic acids. A structure for deoxyribos nucleic acid. *Nature* 171:737–738.

23. Westgren, A. (1972) The Prize in Chemistry. In *Nobel, the Man and His Prizes, 3rd edition*. Odelberg, W. (ed.). Elsevier, New York, pp. 281–385.

24. Todd, A. (1958) Synthesis in the study of nucleotides. In *Les Prix Nobel en 1957*. P. A. Norstedt & Söner, Imprimerie Royale, Stockholm, pp. 119–133.

25. Caspersson, T., Zech, L. and Johansson, C. (1970) Differential banding of alkylating flurochromes in human chromosomes. *Exp. Cell Res.* 60:315–319.

26. Koestler, A. (1964) *The Act of Creation*. Hutchinson & Co., London.

27. McCarty, M. (1985) *The Transforming Principle*. W. W. Norton, New York.

28. Liljestrand, G. (1972) The Prize in Physiology or Medicine. In *Nobel, the Man and His Prizes, 3rd edition*. Odelberg, W. (ed.). Elsevier, New York, pp. 139–278.

29. Caspersson, T. (1948) Introductory speech to the Nobel Prize for Physiology or Medicine 1946. In *Les Prix Nobel en 1946*. P. A. Norstedt & Söner, Imprimerie Royale, Stockholm, pp. 35–38.

30. Caspersson, T. (1959) Introductory speech to the Nobel Prize for Physiology or Medicine 1958. In *Les Prix Nobel en 1958*. P. A. Norstedt & Söner, Imprimerie Royale, Stockholm, pp. 28–31.

31. Berg, P. and Singer, M. (2003) *George Beadle. An Uncommon Farmer. The Emergence of Genetics in the 20th Century*. Cold Spring Harbor Laboratory Press, New York.

32. Beadle, G. W. (1959) Genes and chemical reactions in Neurospora. In *Les Prix Nobel*

en 1958. P. A. Norstedt & Söner, Imprimerie Royale, Stockholm, pp. 147–159.

33. Tatum, E. L. (1959) A case history in biological research. In *Les Prix Nobel en 1958*. P. A. Norstedt & Söner, Imprimerie Royale, Stockholm, pp. 160–169.

34. Watson, J. D. (2007) *Avoid Boring People*. Alfred A. Knopf, New York.

35. Theorell, H. (1960) Introductory speech to the Nobel Prize in Physiology or Medicine 1959. In *Les Prix Nobel en 1959*. P. A. Norstedt & Söner, Imprimerie Royal, Stockholm, pp. 30–33.

36. Ochoa, O. (1960) Enzymatic synthesis of ribonucleic. In *Les Priz Nobel en 1959*. P. A. Norstedt & Söner, Imprimerie Royale, Stockholm, pp. 146–164.

37. Kornberg, A. (1989) *For the Love of Enzymes: The Odyssey of a Biochemist*. Harvard University Press, Cambridge, MA.

38. Ridley, M. (2006) *Francis Crick: Discoverer of the Genetic Code*. Harper Collins, New York.

39. Crick, F. H. C. (1963) On the genetic code. In *Les Prix Nobel en 1962*. P. A. Norstedt & Söner, Imprimerie Royale, Stockholm, pp. 179–187.

40. Watson, J. D. (1963) The involvement of RNA in the synthesis of proteins. In *Les Prix Nobel en 1962*. P. A. Norstedt & Söner, Imprimerie Royale, Stockholm, pp. 155–178.

41. Brenner, S., Jacob, F. and Meselson, M. (1961) An unstable intermediate carrying information from genes to ribosomes for protein synthesis. *Nature* 190:576–581.

42. Nirenberg, M. and Leder, P. (1964) RNA codewords and protein synthesis. The effect of trinucleotides upon the binding of sRNA to ribosomes. *Science* 145:1399–1407.

43. Fleischmann, R. D., Adams, M. D., White, O. *et al.* (1995) Whole-genome random sequencing and assembly of haemophilus influenzae Rd. *Science* 269:496–512.

44. Myers, E. W., Sutton, G. G., Delcher, A. L. *et al.* (2000) A whole-genome assembly of Drosophila. *Science* 287:2196–2204.

45. International Human Genome Consortium (2004) Finishing the euchromatic sequence of the human genome. *Nature* 431:931–945.

46. Levy, S., Sutton, G., Ng, C. P. *et al.* (2007) The diploid genome sequence of an individual human. *PLoS Biol.* 5:3–34.

47. Green, R. E., Krause, J., Briggs, A., W. *et al.* (2010) A draft sequence of the Neanderthal genome. *Science* 328:710–722.

48. Brenner, S. (2003) Nature's gift to science. In *Les Prix Nobel 2002*. Almqvist & Wiksell International, Stockholm, pp. 274–282.

49. Venter, J. C., Remington, K., Heidelberg, J. F. *et al.* (2004) Environmental genome shotgun sequencing of the Sargasso Sea. *Science* 304:66–74.

50. Yooseph, S., Sutton, G., Rusch, D. B. *et al.* (2007) The Sorcerer II global ocean sampling expedition: Expanding the universe of protein families. *PLoS Biology* 5:432–466.

51. Cello, J., Paul, A. V. and Wimmer, E. (2002) Chemical synthesis of poliovirus cDNA: Generation of infectious virus in the absence of natural template. *Science* 297:1016–1018.

52. Tumpey, T. M., Basler, C. F., Aguilar, P. V. *et al.* (2005) Characterization of the reconstructed 1918 Spanish influenza pandemic virus. *Science* 310:77–80.

53. Smith, H. O., Hutchinson III, C. A., Pfannkoch, C. and Venter, J. C. (2003)

Generating a synthetic genome by whole-genome assembly: φX174 bacteriophage from synthetic oligonucleotides. *Proc. Natl. Acad. Sci. USA* 100:15440–15445.

54. Gibson, D. G., Glass, J. I., Lartigue, C. *et al.* (2010). Creation of a bacterial cell controlled only by a chemically synthesized genome. *Science* 329:52–56.

55. Gesteland, R. F., Cech, T. R., Atkins, J. F. (eds.) (2005) *The RNA World, 3rd edition.* Cold Spring Harbor Laboratory Press.

56. Joyce, G. F. (2002) The antiquity of RNA-based evolution. *Nature* 418:214–221.

57. ENCODE Project Consortium (2007) Identification and analysis of functional elements in 1% of the human genome by the ENCODE pilot project. *Nature* 447:799–816.

Chapter 8

1. Prusiner, S. B. (1982) Novel proteinaceous infectious particles cause scrapie. *Science* 216:195–206.

2. Farquhar, J. and Gajdusek, D. G. (eds.) (1981) *Kuru. Early Letters and Field-notes from the Collection of D. Carleton Gajdusek.* Raven Press, New York.

3. Gajdusek, C. (1977) *Les Prix Nobel en 1976.* P. A. Norstedt & Söner, Imprimerie Royale, Stockholm, pp. 161–166.

4. Klein, G. (1997) *Live Now.* Prometheus, New York, pp. 85–164.

5. Rhodes, R. (1997) *Deadly Feasts: Tracking the Secrets of a Terrifying New Plague.* Simon & Schuster, New York.

6. Redfield Jamison, K. (2004) *Exuberance.* Random House, New York, pp. 206–211.

7. Anderson, W. (2008) *The Collectors of Lost Souls. Turning Kuru Scientists into Whitemen.* Johns Hopkins University Press, Baltimore.

8. de Kruif, P. (1926) *Microbe Hunters.* Harcourt, New York.

9. Zuckerman, H. (1996) *Scientific Elite. Nobel Laureates in the United States.* New edition with a new introduction. Transaction Publishers, New Brunswick.

10. Watson, J. D. (2007) *Avoid Boring People.* Alfred A. Knopf, New York.

11. Gajdusek, C. and Zigas, V. (1957) Degenerative disease of the central nervous system in New Guinea: The endemic occurrence of "kuru" in the native population. *New Engl. J. Med.* 257:974–978.

12. Hadlow, W. J. (1959) Scrapie and kuru. *Lancet* 2:289–290.

13. Gajdusek, D. C., Gibbs, C. J., Jr. and Alpers, M. (1966) Experimental transmission of a kuru-like syndrome to chimpanzees. *Nature* 209:794–796.

14. Gibbs, C. J., Jr., Gajdusek, D. C., Asher, D. M. *et al.* (1968) Creutzfeldt-Jakob disease (spongiform encephalopathy): Transmission to the chimpanzee. *Science* 161:388–389.

15. Glasse, R. M. (1967) Cannibalism in the kuru region of New Guinea. *Trans. N.Y. Acad. Sci.* 29:748–754.

16. Collinge, J., Whitfield, J., McKintosh E. *et al.* (2006) Kuru in the 21st century — an acquired human prion disease with very long incubation periods. *Lancet* 367:2068–2074.

17. Brown, P., Brandel, J.-P., Preece, M. and Sato, T. (2006) Iatrogenic Creutzfeldt-Jakob disease: The waning of an era. *Neurology* 67:389–393.

18. Dorsey, K. A., Zou, S., Schonberger, L. B. *et al.* (2009) Lack of evidence of

transfusion transmission of Creutzfeldt-Jakob disease in a U.S. surveillance study. *Transfusion* 49:977–984.

19. Houston, F., McCutcheon, S., Goldmann, W. *et al.* (2008) Prion diseases are effectively transmitted by blood transfusion in sheep. *Blood* 112:4739–4745.

20. Gajdusek, D. C. (1977) Unconventional viruses and the origin and disappearance of kuru. In *Les Prix Nobel en 1976*. P. A. Norstedt & Söner, Imprimerie Royale, Stockholm, pp. 67–216.

21. Alper, T., Cramp, W. A., Haig, D. A. and Clarke, M. C. (1967) Does the agent of scrapie replicate without nucleic acid? *Nature* 214:764–766.

22. Griffith, J. S. (1967) Self-replication and scrapie. *Nature* 215:1043–1044.

23. Prusiner, S. B. (1998) Prions. In *Les Prix Nobel en 1997*. Almqvist & Wiksell International, Stockholm, pp. 268–323.

24. Prusiner, S. B., Groth, D. F., Bolton, D. C. *et al.* (1984) Purification and structural studies of a major scrapie prion protein. *Cell* 38:127–134.

25. Buchan, J. (2003) *Crowded with Genius. The Scottish Enlightenment: Edinburgh's Moment of the Mind*. HarperCollins, New York.

26. Oesch, B., Westaway, D., Wälchli, M. *et al.* (1986) A cellular gene encodes scrapie PrP 27–30 protein. *Cell* 46:417–428.

27. Büeler, H., Fischer, M., Lang, Y. *et al.* (1992) Normal development and behaviour of mice lacking the neuronal cell-surface PrP protein. *Nature* 356:577–582.

28. Büeler, H., Aguzzi, A., Sailer, A. *et al.* (1993) Mice devoid of PrP are resistant to scrapie. *Cell* 73:1339–1347.

29. Prusiner, S. B., Groth, D., Serban, A. *et al.* (1993) Ablation of the prion protein (PrP) gene in mice prevents scrapie and facilitates production of anti-PrP antibodies. *Proc. Natl. Acad. Sci. USA* 90:10608–10612.

30. Colby, D. W., Giles, K., Wille, H. *et al.* (2009) Design and construction of diverse mammalian prion strains. *Proc. Natl. Acad. Sci. USA* 106:20417–20422.

31. Chesebro, B., Race, B., Meade-Whit, K. *et al.* (2010) Fatal transmissible amyloid encephalopathy: A new type of prion disease associated with lack of prion membrane anchoring. *PLoS Pathogens* 6:1–14.

32. National hormone and pituitary program: Information for people treated with pituitary human growth hormone (summary). U.S. Department of Health and Human Services, National Institutes of Health. August 2009. www.endocrine. niddk.nih.gov

33. Will, R. G., Ironside, J. W., Zeidler, M. *et al.* (1996) A new variant of Creutzfeldt-Jakob disease in the U.K. *Lancet* 347:921–925.

34. The National Creutzfeldt-Jakob Disease Surveillance Unit (NCJDSU), University of Edinburgh, U.K., www.cjd.ed.ac.uk. (2009).

35. Llewelyn, C. A., Hewitt, R. E., Knight, R. S. G. *et al.* (2004) Possible transmission of variant Creutzfeldt-Jakob disease by blood transfusion. *Lancet* 363:417–421.

36. Donne, D. G., Viles, J. H., Groth, D. *et al.* (1997) Structure of the recombinant full-length hamster protein protein PrP (29–231): The N terminus is highly flexible. *Proc. Natl. Acad. Sci. USA* 94:7279–7282.

37. Riek, R., Hornemann, S., Wider, G. *et al.* (1997) NMR characterization of the full-length recombinant murine prion protein, mPrP (23–231). *FEBS Lett.* 413:282–288.

38. Prusiner, S. (2001) Shattuck lecture — Neurodegenerative diseases and prions.

New Engl. J. Med. 344:1516–1526.

39. Kyle, R. A. (2001) Amyloidosis: A convoluted story. *Brit. J. Haem.* 114:529–538.
40. Aguzzi, A. (2009) Beyond the prion principle. *Nature* 459:924–925.
41. Wickner, R. B., Edskes, H. K., Shewmaker, F. and Nakayashiki, T. (2007) Prions of fungi: Inherited structures and biological roles. *Nat. Rev. Microbiol.* 5:611–618.
42. Alberti, S., Halfmann, R., King, O. *et al.* (2009) A systematic survey identifies prions and illuminates sequence features of prionogenic proteins. *Cell* 137:146–158.
43. Johnson, P. (1988) *Intellectuals*. Harper & Row, New York.
44. Redfield Jamison, K. (1993) *Touched with Fire*. Simon & Schuster, New York.
45. Brown, P. (2009) Daniel Carleton Gajdusek (1923–2008). *Neurology* 72:1204.
46. Goudsmit, J. (2009) Daniel Carleton Gajdusek (1923–2008). *Nature* 457:394.
47. Prusiner, S. B. and McCarty, M. (2006) Discovering DNA encodes heredity and prions are infectious proteins. *Ann. Rev. Genet.* 49:25–45.

Name Index

Brown, P., 254, 254f, 277
Bruynoghe, G., 135
Buchner, E., 53, 53f
Buchner, H., 53
Burnet, F. M., 63, 71, 71f, 87, 91, 114, 134, 155, 185, 249, 251, 262, 275

C

Cairns, J., 238
Campbell, W. R., 166
Cannon, W. B., 45
Capecchi, M. R., 263
Carl XVI Gustaf, 257f
Carlson, A. J., 139
Carrel, A., 72, 117, 146t, 150–154, 152f, 154f
Carroll, J., 67, 103
Carson, R., 182
Casals, J., 106
Caspersson, T., 72, 77, 77t, 89, 209, 209f, 210–212, 214, 217, 219, 222
Cech, T. R., 224t, 241, 241f
Chain, E. B., 56, 147t, 176–178, 178f, 190
Chargraff, E., 203, 215
Chase, M., 84, 85, 204, 226
Cohen, S. S., 74, 83
Collip, J. B., 162, 166, 167, 167f, 168
Cori, C. F., 51, 223
Cori, G. T., 51, 223
Cournand, A. F., 155
Cox, H., 125, 135, 138, 142f
Crafoord, H., 34
Crick, F. H. C., 194–196, 197f, 203, 206, 207, 219, 224t, 225, 227, 228f, 229–235, 230f, 241, 242, 274
Crile, G. W., 162
Curie, M., 26, 49, 120, 231
Curie, P., 49, 231
Cusack, J., 60

D

Dale, H. H., 169, 170f, 171, 215
Dalén, G., 13
Dam, H. C. P., 147t, 173, 174, 175, 175f, 186, 190
Darwin, C. R., 44, 194, 195, 217
de Kruif, P., 247f, 248
de Lesseps, F., 100
Delbrück, M., 70t, 72, 74, 80, 82, 83, 84f, 85, 248
Descartes, R., 96

d'Herelle, F., 82, 83, 168
Dickinson, A., 262
Dingle, J. H., 134
Dobzhansky, T., 95
Dodge, A., 60
Doherty, P. C., 70t, 73
Doisy, E. A., 147t, 173–175, 175f, 186, 190
Domagk, G., 56, 56f, 57
Donnall, T. E., 155
du Vigneaud, V., 136, 137, 186, 190
Dubos, R. J., 91, 216
Dulbecco, R., 70t, 73, 89, 138, 224t, 229
Dunant, J. H., 25

E

Edelman, G. M., 233
Ehrlich, P., 150, 157
Eigen, M., 275
Einstein, A., 12
Eklund, C., 261
Ellerman, D. C., 72
Ellis, E., 83
Embden, G., 158, 159, 160, 161
Emerson, R. A., 218
Enders, J. F., 69t, 71, 89, 119, 123–125, 128–139, 137f, 141, 145, 146t, 249
Engström, A., 221
Epstein, D., 275
Ericsson, J., 4
Erlanger, J., 51, 177
Ernberg, I., 276
Evans, M. J., 263

F

Fankuchen, I., 78
Feldt, K.-O., 28
Feller, A., 130
Feynman, R., 63
Fibiger, J., 115, 157
Fieser, L. F., 173, 174
Finlay, C., 100
Finsen, N. R., 148, 149, 150
Fire, A. Z., 70t, 73, 224t, 242
Fischer, E., 199, 200, 205
Fischer, G. A. V., 179, 180, 181
Fitzgerald, J. G., 165
Fleming, A., 55, 56, 147t, 176, 177, 178, 178f
Flemming, W., 198
Flexner, S., 103, 104, 129, 133, 139, 152, 154
Florey, H. A., 56, 147t, 176, 177, 178, 178f, 190

Forssmann, W., 155
Fox, J., 106
Fraenkel-Conrat, H. L., 74, 80, 81, 92, 204, 207
Francis, T., 139
Franklin, R., 69
Fredga, A., 205, 206, 206f
Freneau, P., 99
Friberg, U., 220f, 221
Frosch, P., 67
Frost, R., 64
Funk, C., 175
Furchgott, R., 40, 41f

G

Gajdusek, D. C., 57, 69t, 73, 113, 245–261,
   247f, 257f, 273, 275–278
Gard, S., 66, 66f, 70–72, 74, 85–90, 106, 107,
   109–116, 119, 120, 122, 123, 129–136, 132f,
   139–142, 188, 215
Gasser, H. S., 106, 154, 177, 213, 214
Gebhardt, L. P., 129
Geigy, J. R., 179, 180
George V, 101
Gibbs, Jr., C. J., 253, 253f
Gierer, A., 81, 92, 204
Gilbert, W., 224t
Gilot, F., 143
Golgi, C., 148
Goodpasture, E. W., 87
Gorgas, W. C., 101
Göring, H., 154
Goudsmit, J., 278
Grassi, B., 248
Griffith, F., 213
Grubb, R., 140
Guillemin, R., 186
Gustaf VI Adolf, 98f, 197f

H

Haagen, E., 111, 113
Hadlow, W., 251, 251f, 252
Hagedorn, H. C., 169
Hamburger, V., 248
Hammarsten, E., 77, 77t, 79, 160, 174, 184,
   185, 209, 209f, 210, 214, 215, 219, 220
Hammarsten, O., 200, 209
Harrison, R. G., 151
Hastings, B., 220
Hedén, C.-G., 211
Hedrén, G., 103

Heeger, A. J., 54, 54f
Heidelberger, M., 213, 215
Heine, J., 127
Hellerström, S., 177
Hellström, J., 116, 136
Hemingway, E., 168
Hench, P. S., 114, 115, 146t, 183–185, 185f, 191
Henry, J., 61
Henschen, F., 16, 72, 77t, 109, 115, 217
Herriott, R., 204
Hershey, A. D., 70t, 72, 74, 81, 82, 83, 84, 84f,
   85, 92, 204, 226
Hertwig, O., 199
Hewish, A., 50, 50f, 51
Heyrovsky, J., 220
Hill, A. V., 146t, 158, 159, 159f, 160, 190
Hilleman, M., 118, 118f
Hirst, G.K., 135
Hitler, A., 57, 154
Hoffmann, R., 63
Holley, R. W., 224t, 231, 232f
Hoppe-Seyler, F., 198, 199, 200
Horne, R. W., 225
Horstmann, D., 135
Horvitz, H. R., 227f, 229
Hotchkiss, R., 204, 208, 214
Hoyle, L., 88
Hoyt, G. H., 160
Hüfner, G., 201
Huggins, C., 221
Hürtle, K., 158
Hutchinson, C. A., 239
Huxley, T. H., 44
Hydén, H. V., 211, 212

I

Ignarro, L., 40, 41f
Innes, J. R. M., 251
Ivanovsky, D. I., 65

J

Jacob, F., 68f, 70t, 72, 188
Jacobaeus, H. C., 162, 163, 164
Jangfeldt, B., 11
Jefferson, T., 100
Jenner, E., 55, 116, 248
Johannsen, W., 216
Johansson, J. E., 8, 23, 145, 145f, 156, 158,
   160, 191
Johnson, K., 260

Melville, H., 259, 275
Mendel, J. G., 195, 216, 217
Merton, R. K., 46, 46f, 59, 63, 64
Meyerhof, O. F., 72, 146t, 158, 159, 159f, 160, 222
Meynell, W., 45
Miescher, F., 198f
Milstein, C., 153
Minot, G. R., 160
Mirsky, A., 204, 211
Mischer, F., 197, 198, 199, 200, 201, 210
Mittag-Leffler, G., 14
Monod, J., 68f, 70t, 72, 188, 195
Monroe, J., 100
Montagnier, L., 69t, 73, 118, 192
Morgan, T. H., 83, 191, 217, 217t
Morison, R. S., 106
Mörner, K. A. H., 200, 201, 201f, 233
Morrison, J. H., 113
Mozart, W. A., 275
Muller, H. J., 191, 217, 217t
Müller, P. H., 110, 146t, 179, 180, 181, 181f, 190
Mullis, K. B., 224t
Murad, F., 40, 41f
Murphy, W. P., 160
Murray, J. E., 72, 155
Mussolini, B., 154
Myrbäck, K. D. R., 219, 220

N

Nathans, D., 30, 70t, 72, 224t
Nencki, M., 9
Neuberg, K., 157
Neubuerger, K. T., 139
Newton, I., 48
Nicolle, C. J. H., 179
Nirenberg, M. W., 224t, 231, 232f
Nobel, A., 2, 3f, 4, 5, 6, 8, 9, 10, 11, 12, 14, 16, 18f, 28, 32, 33, 34, 37f, 38, 39, 40, 40f, 82, 97, 99, 118, 119, 123, 125, 144, 148, 160, 161, 184, 193
Nobel, C. A., 2, 7f, 8
Nobel, E., 4, 14, 15, 16
Nobel, I., 2, 5
Nobel, L., 4, 11
Nobel, R., 4, 11
Nobelius, P. O., 2
Noble, R. E., 101
Noguchi, H., 103, 104, 107, 153, 248
Nordenskiöld, A., 5, 6

Northrop, J. H., 68, 68t, 74, 74f, 76, 77, 79, 83, 130, 190, 207, 208
Norton, T.W., 108

O

Ochoa, S., 147t, 196, 219, 220, 221, 222, 222f, 223, 224t, 237
O'Connor, B., 127
Olitsky, P., 129, 133
Orgel, L., 229
Osborn, L. A., 139
Oscar II, 14

P

Papanicolaou, G. N., 221
Pasteur, L., 47, 52, 61, 64, 67, 117, 120, 215
Paul, J. R., 134
Pauling, L., 248
Pavlov, I. P., 9, 148
Peebles, J., 49
Pelouze, T.-J., 4
Penzias, A., 49, 50, 50f
Perutz, M. F., 230, 230f, 235
Pettersson, A., 164
Pettersson, R. F., 71, 246
Pfeiffer, R. F. J., 87
Phipps, J., 55
Pirie, N. W., 76, 76t, 78, 79, 91
Porter, R. R., 233
Powell, J. H., 99
Prusiner, S. B., 57, 69t, 73, 125, 134, 143, 245, 246, 260, 261–266, 261f, 278, 278f, 279, 279f

R

Ramakrishnan, V., 224t, 231
Ramel, S., 28, 29, 33
Ramón y Cajal, S., 148
Redfield Jamison, K., 246, 275, 276
Reed, W., 67, 100, 101, 101f, 103
Reichard, P., 209, 210
Reichstein, T., 114, 115, 146t, 183, 184, 185, 185f, 190, 191
Richards, D., 155
Richet, C., 150
Rivers, T. M., 86
Robbins, F. C., 69t, 71, 89, 119, 124, 125, 128, 130, 131, 133–139, 137f, 141, 143, 145, 146, 146t, 189
Roberts, R. J., 70t, 73, 224t, 241

Roberts, R. M., 47
Robscheit-Robbins, F., 52
Rockefeller, Jr., J. D., 113
Roentgen, W. C., 48, 48f, 49
Roosevelt, F., 127
Ross, R., 15, 101, 148, 149, 248
Rous, P., 32, 69t, 72, 191
Roux, E., 248
Rudbeck, O., 2
Rusk, H.A., 139
Russell, F., 104
Ryle, M., 51

S

Sabin, A., 109, 116, 119, 125, 129, 133, 138, 140, 141, 142f
Salk, J., 119, 125, 131, 135, 137, 138, 139, 140, 141, 142, 142f, 143
Saltin, B., 160
Salvat Navarro, A., 114
Samuelsson, B. I., 212
Sanger, F., 26, 224t, 234, 234f, 235, 238
Sarett, L. H., 184
Sarpyener, A., 139
Sauerbruch, F., 155
Sawyer, W. A., 104, 112, 113, 114
Schafer, S., 162
Schally, A. V., 186
Schliemann, H., 45
Schmitz, E., 158
Schramm, G., 74, 80, 81, 92, 204
Sellards, A. W., 107
Sharp, P. A., 70t, 73, 224t, 241
Shawn, W., 182
Sherrington, C. S., 32, 161, 191
Shirakawa, H., 53, 54, 54f
Shope, R. E., 87, 88
Siegbahn, 231
Sigurdsson, B., 252
Sjögren, A., 5
Sjöqvist, J., 156, 158, 159, 162, 163, 166
Smadel, J., 249, 252, 253
Smith, G., 13
Smith, H. O., 30, 70t, 72, 224t, 239, 240, 240f
Smith, M., 224t
Smith, T., 248
Smith, W., 87
Smithies, O., 263
Sobrero, A., 4, 39
Sohlman, M., 17

Sohlman, R., 9, 16
Solly, E., 44
Spemann, H., 152, 191
Stadtman, E., 260
Stanley, W. M., 68, 68t, 74, 74f, 75, 76, 77, 78, 79, 80, 82, 83, 88, 89, 90, 91, 92, 93, 190, 196, 202, 205, 208, 217
Steinbeck, J., 230f
Steitz, T. A., 224t, 231
Stewart, J., 104
Stokes, A., 103, 104, 107
Strömbeck, J. P., 173
Sulston, J. E., 227f, 229
Sumner, J. B., 76, 77
Sundberg, C., 103
Svartz, N., 176, 177, 184
Svedberg, T., 75, 75f, 76, 76t, 77t, 78, 88, 131, 175, 205, 209
Svedmyr, A., 258, 259
Svedmyr, B., 258, 259

T

Tamm, G., 16
Tatum, E. L., 63, 72, 186, 187, 188, 196, 214, 217, 217t, 218, 218f, 241
Taub, A., 80
Taussig, H. B., 155
Temin, H. M., 70t, 73, 224t, 229
Theiler, L., 122
Theiler, M., 69t, 71, 88, 97, 98, 98f, 102, 106–122, 122f, 131, 143, 146, 147t, 185, 189
Theorell, H., 137, 183, 190, 211, 213, 220f, 221, 222, 243
Thomas, V. T., 155
Thomson, 231
Timofeeff-Ressovsky, N., 82
Tiselius, A. W. K., 75, 75f, 76, 76t, 80, 81, 88, 131, 205, 207, 208
Tjio, J. H., 212
Todd, A. R., 196, 205, 206, 207, 222
Tonegawa, S., 233
Toynbee, A. J., 108
Twort, F., 82, 83

U

Ullgren, C., 5
Urey, H. C., 74
Uvnäs, B., 256

Subject Index

A

acquired immunodeficiency syndrome (AIDS), 73

Act of Creation, The (Koestler), 211

ACTH. *See* adrenocorticotropic hormone

adrenal hormones, 157, 183–184

adrenocorticotropic hormone (ACTH), 183–184

Africa, yellow fever and, 101

AIDS. *See* acquired immunodeficiency syndrome

Alzheimer's disease, 272

amino acids

    cystein, 234

    genetics and, 232–233. *See also* genetics

    insulin and, 234. *See also* insulin

    methionine, 232

    nucleic acids and. *See* nucleic acids

    phenylalanine, 231

    proteins and, 203, 232–233, 234. *See also* proteins

    PrP and, 262

    TMV and, 80

    yellow fever virus and, 122

amyloid diseases, 272

anaemia, 160

antibiotics

    Domagk and, 56–57

    Fleming and, 176–178

    monolayer cultures and, 117

    penicillin. *See* penicillin

    prontosil, 56

    streptomycin, 116, 134

    tuberculosis and, 116

    See also bacteria

antibodies, 153, 233

    antigens and, 58, 59

    Blumberg and, 57–59

    Carrel–Noguchi experiment, 153

    catalytic, 233

    chemical structure of, 233

    CNS disease and, 125

    immunity and. *See* immune system

    maternal, 125, 127

    monoclonal, 153

    poliovirus and, 125, 127

    PrP and, 263

    TMV and, 80, 82

    vaccination and. *See* vaccines

    yellow fever and, 105, 111–112

Arrowsmith (Lewis), 168

aspirin, 186

astrophysics, 51

Australia antigen, 57

Avoid Boring People (Watson), 248

B

bacteria

    antibiotics. *See* antibiotics

    bacteriophages and. *See* bacteriophages

    discovery of, 64

    genetic material in, 94, 187, 235

    Nobel on, 8

    sizes of, 66

    species of, 94

    spirochetes and, 103

    transformation and, 203, 213–214, 215

    viruses and, 67. *See also* virology

    See also specific types, diseases

genetics
  prophage and, 187
  proteins and, 208, 243. *See also* proteins,
    specific topics
  recombination and, 188
  regulation systems, 242
  reverse transcriptase, 73
  RNA and. *See* ribonucleic acid
  sequencing technologies, 94
  splicing and, 73
  split genes, 70t, 241
  strain variations, 278
  transduction and, 187, 188
  viruses and. *See also* virology; *specific*
    types, topics
  yeast cells and, 189
Germany, 57. *See also specific awards, topics*
Great Britain, 267. *See also specific awards, topics*
growth hormone, 264

H

haemoglobin, 205
Haiti, 100
heart surgery, 155
hepatitis virus, 58–59, 113, 117, 245, 256,
  268, 270
HIV. *See* human immunodeficiency virus
Högskola (Sweden), 13
homeostasis, 243
hormones, 205
  adrenal, 157, 183–184
  corticosteroids, 183–186, 192
  growth hormone, 264
  insulin. *See* insulin
  polypeptide, 186
human immunodeficiency virus (HIV), 73,
  118, 192, 270
Huntington's disease, 272
hyaluronic acid, 58

I

iatrogenic disease, 269
immune system, 189
immunity, 149, 157
  bacteria and. *See* bacteria
  cell mediation and, 70t
  clonal selection, 249
  immunological differences, 269

vaccines and, 105. *See* vaccines
virology and, 63, 71. *See also* virology;
  specific topics
immunological amyloidosis, 272
infection, principle of, 245
influenza, 118, 130
  Gard and, 87–88
  RNA genome, 238
  virology and, 87–88
insecticides, 179–183
Institute of Man (US), 153
insulin, 157
  amino acids in, 234
  Banting and, 161–168
  Best and, 166, 167, 168, 171
  Dale nomination, 170–171
  discovery of, 161–172
  first treatment, 162
  insulin adventure, 169
  Krogh and, 168–169
  MacLeod and, 161–168
  nominations for, 162
  structure of, 234
  von Euler on, 172
Intellectuals (Johnson), 274
interfering ribonucleic acid (RNAi), 73
invention, 13, 33. *See also* discovery

J

jaundice, 57, 102

K

Karolinska Institute, 6, 13, 16, 22, 24, 75
  archive materials, 98, 125, 193. *See also*
    specific nominations, awards
  awards to organizations, 55
  C. A. Nobel Endowment, 8
  College of Teachers, 136
  E. Nobel and, 15
  Faculty of, 19, 98, 189
  Institute for Cell Research, 210
  interpretation of will, 144
  Nobel Committees and, 19, 20, 220. *See*
    also Nobel Committees
  Nobel will and, 9, 10, 11, 17, 28, 118, 144,
    157, 189
  nucleic acids and, 208
  selection process, 193

serendipity
  Becquerel and, 49
  Big Bang theory and, 49
  Cannon on, 45
  chemistry and, 52
  Fleming and, 55–56
  Henry on, 61
  history of term, 44–47
  Merton on, 46, 64
  origin of term, 41
  penicillin and, 55
  physics and, 48
  polymers and, 53
  pulsars and, 51
  Roentgen and, 48
  Walpole and, 41
  X-rays and, 48
Serendipity (film), 60
serum therapy, 149
Silent Spring (Carson), 182
smallpox, 54, 116, 117, 124
space flights, 189
species barrier, prions and, 263, 267
spiritualism, 153
spirochetes, 103
streptomycin, 116
Study of History, A (Toynbee), 108
Suez Canal, 100
sugar, cleavage of, 157
sulphonamides, 56
surgery, Carrel and, 152
Sweden, polio and, 126f
Swedish Academy, 13
Swedish Club (Paris), 11, 12
syphilis, 103, 150

T

T-even phages, 92
Thursday Club, 234
thyroid gland, 155, 157
tissue culture techniques, 120, 151–152
tobacco mosaic virus (TMV), 78–81, 190
  crystallization of, 92
  Prizes and, 68–69, 68t
  RNA from, 92, 204, 207
  shape of, 92, 93
  virology and, 74, 89, 93
transduction, 86, 187, 188
transformation, 203
Travels and Adventures of Serendipity
  (Merton/Barber), 46, 59, 63
tropisms, 105

trypsin, 117
tuberculosis, 116, 150
typhus, 179–180

U

United States
  American laureates, 26–28, 27t, 107
  democratization of science, 36
  Louisiana Purchase and, 100
  Panama Canal and, 101
  Philadelphia plague, 99
  polio and, 126f
  yellow fever and, 99–103
university research programs, 36

V

vaccines, 55, 99, 105
  attenuated strains and, 109
  developing countries and, 118
  development of, 105
  discovery and, 120
  DNA virus infections and, 118
  embryonated eggs, 117, 120
  enteric infections, 118
  hepatitis and, 58–59, 113, 117, 245, 256,
    268, 270
  history of, 116
  immunity and, 105, 120. *See also*
    immunity
  influenza and, 118
  laboratory infections, 112
  measles and, 117, 118, 130, 137
  mumps and, 117, 130
  polio and, 117, 124, 125, 128. *See also*
    polio
  rabies and, 117
  recombinant DNA methods, 59
  Rockefeller Foundation and, 113
  roller tube culture system and, 130
  smallpox, 54, 116, 117, 124
  tests of, 111
  tissue culture techniques, 130
  trypsin technique, 137
  yellow fever and, 105
  See also specific diseases, topics, awards
varicella virus, 130, 143
vascular anastomosis, 150–151
vertical infection and, 268
Viagra, 40
virology, 57–58, 65
  animal experimentation and, 253

virology
animal viruses and, 86–90
antibodies and, 105
antigens and, 57–58
average incubation time, 261
bacteria and, 90
bacteriophages and, 82–86
bioterrorism and, 238
Blumberg and, 57–58
Burnet and, 71
cancer and, 85
cannibalism and, 268–269
central nervous system and, 124
concept of virus, 66–96
counterfactual history of, 92–93
cytopathic effects, 137
developments in, 93–95
discovery and, 119
DNA and, 84
eclipse phase, 84
egg system, 88, 117, 120
electron microscopy and, 89, 225, 259
enzymes and, 70t, 76, 79, 83, 221, 261
epidemics and, 97. *See* epidemics
evolution and, 95. *See also* genetics
first definition of, 67–68
first Prize in, 74–81
flavivirus and, 105
Gard and, 66, 70, 71, 88, 90
genetics and, 73, 82–86, 91. *See* genetics
heredity and, 85. *See* genetics
Hershey–Chase experiment, 92
high-tech cannibalism, 269
history of, 92–93, 116
immunity. *See* immunity
immunological tolerance, 262
immunology and, 71
infectious viruses, 94. *See specific types,*
topics
influenza and, 87–88
interference, 84
"knock-out" technique, 263
larger viruses, 89
life and, 95
lysogeny and, 84, 85
lytic cycle, 91
molecular science and, 68, 91, 93
monolayer cultures and, 89, 117
mosaic patterns, 67. *See also* TMV
nature of viruses and, 86–90
neurotropic properties, 105, 111, 120

Noah's approach, 253
Nobel Prizes and, 68–73, 69t, 70t. *See*
specific awards, topics
nucleic acids and, 78, 80, 81, 89, 91, 92
plants and, 76, 90, 93. *See also* TMV
plaque technique, 89
polio. *See* polio
prions. *See* prions
Prizes and. *See specific awards, topics*
prophages and, 85
recombination and, 84
replication and, 105
RNA and, 80, 124, 237
Rockefeller Foundation and, 108
slow viruses, 245, 252
species barrier, 263, 267
Stanley and, 74–81
sub-cellular infectious agents, 245
symmetry of particles, 225
synthesis of viruses, 238
Theiler and, 119
tissue culture techniques, 117, 120
tissue cultures, 111
TMV and. *See* tobacco mosaic virus
transduction, 84, 85
trypsin and, 117
tumor virus, 72
vaccines. *See* vaccines, *specific types, topics*
vertical infection and, 268
virus, concept of, 66–96
virus particle and, 237
virus structure, 89
viscerotropic properties, 105, 111, 120
yellow fever and, 88. *See* yellow fever
See also specific topics
Viruses and man (Burnet), 91
vitalism, theory of, 52–53, 90
vitamins, 160, 173–175, 186, 192, 205
Voltaire, 44

W

West Indies, yellow fever and, 100
West Nile virus, 121
World Health Organization (WHO), 124, 182

X

X-ray crystallography, 230–234

Y

yellow fever, 67, 96
   Africa and, 101–102
   antibodies and, 111
   Asia and, 102
   Asibi strain, 104, 111, 120, 121
   bacteria and, 103
   Brazil and, 101, 112
   Cuba and, 100
   DDT and, 180
   deaths of researchers, 104
   epidemiology of, 99–103
   Equador and, 101
   Europe and, 102
   field studies, 102
   flaviviruses and, 105
   French strain, 107
   in mice, 108, 110–111
   in monkeys, 102–104, 107–111
   jaundice and, 102, 112, 113
   jungle fever and, 102
   laboratory infections, 104, 108, 112
   malaria and, 101
   mosquito vectors, 100, 107
   Napoleon and, 100
   neurotropic properties, 111
   Noguchi and, 104, 107
   origin of, 102
   Panama and, 101
   present-day classification, 105
   proteins in, 105
   Rockefeller Foundation and, 104, 122
   Sabin and, 109
   sanitation and, 101
   17D strain, 109, 111, 114, 116, 121
   spread of, 102
   stability of virus, 106
   Theiler and, 97–123
   ultrafiltration and, 103
   urban form, 102
   US and, 99–103
   vaccine, 105, 114
   viscerotropic properties, 105, 111
   vomito negro and, 102
   West Indies and, 100
   WHO guidelines, 112
   WWII and, 113
   Yellow Fever Commission, 101
   Yellow Jack, 102

Z

Zadig (Voltaire), 44, 64